FLORA OF TROPICAL EAST AFRICA

POLYGONACEAE

R. A. Graham

Herbs, climbers, shrubs, or trees (not in tropical East Africa). Branches sometimes tendrilous. Leaves usually alternate, often lush, sessile or stalked, usually dilated into an amplexicaul cup (ocrea) at the junction with the stem. Flowers actinomorphic ; hermaphrodite, polygamous, dioecious, or monoecious ; borne in many- or few-flowered fascicles, pedicellate, subtended or not by ocreiform bracts with or without bracteoles. Inflorescence capitate, racemose, or ± umbellate, often branched and paniculate. Petals 0. Perianth inferior, calyciform with 3–6 free, sometimes petaloid lobes (tepals) arranged in 1 or 2 series, often accrescent and then sometimes armed with spines, prickles or teeth. Stamens 5–9 in E. African spp., usually all fertile, inserted in 1 or 2 series at or near the base of the perianth ; filaments filiform throughout or basally dilated, free or conjoined at the base ; anthers 2-celled, longitudinally dehiscent. Styles 2–3, free or connate below, the flowers sometimes heterostylous ; stigmas capitate, dilated, fimbriate or penicillate. Ovary 1-locular, superior, sessile. Mature fruit nut-like, fusiform, lenticular, or acutely or obscurely trigonous, unarmed or armed with prickles or (*Harpagocarpus*) barbed setae. Ovule solitary, basal, sessile or stalked. Seed with abundant endosperm.

A world-wide family, most abundant in temperate regions. Two genera are confined to Africa and Madagascar. Certain species have value as crops, as vegetables, or as medicines.

Petioles with basal ocreae ; tendrils not produced :
  Flowers monoecious ; male flowers with 5–6 tepals ; female flowers with 6 tepals in two series of 3 each ; female perianth accrescent, the outer 3 tepals conjoined and bearing stiff, spreading spines at their ends . . . . . . 1. **Emex**
  Flowers hermaphrodite (but dioecious in *Rumex acetosella*) :
    Nuts with two rows of stellately barbed setae down the angles ; perianth somewhat accrescent ; weak, climbing herb . . . . . 2. **Harpagocarpus**
    Nuts unarmed :
      Flowers 3-merous ; inner 3 tepals accrescent, with entire, dentate, or wavy edges . . 3. **Rumex**
      Flowers 4–5-merous ; tepals not accrescent :
        Nuts not exceeding the perianth ; racemes capitate or spiciform-racemose . . 4. **Polygonum**
        Nuts exceeding the perianth ; racemes usually umbellate . . . . . . 5. **Fagopyrum**
  Flowers polygamous, heterostylous ; perianth tube accrescent, prickly, toothed, or neither . . 6. **Oxygonum**
Petioles basally amplexicaul but without ocreae ; tendrilous, climbing shrub or creeper . . . 7. **Antigonon**

FIG. 1.   *EMEX AUSTRALIS*—**1,** flowering branch. showing leaves, fruit and ♂ flowers, × 2/3 ; **2,** fruit, × 3 ; *E. SPINOSUS*—**3,** fruit, × 3 ; *HARPAGOCARPUS SNOWDENII*—**4,** flowering branch, with one lower leaf, × 2/3 ; **5,** fruit, × 3 ; **6,** barbed seta, × 12.

1. EMEX

Neck., Elem. Bot. 2 : 214 (1790), *nom. conserv.*

Monoecious herbs, probably annual. Male flowers with 5–6 tepals, free nearly to the base, spreading, subequal. Stamens 4–6. Female flowers urceolate, with an accrescent, ovoid tube ; tepals 6, arranged in two series, the inner 3 accrescent and erect, the outer 3 accrescent, conjoined, and forming very hard, ± spreading spines at the tips. Styles 3, terminally dilated and fimbriate. Nut acutely trigonous, free within the hardened fruiting perianth.

A small genus comprising two similar but easily distinguished species, native in the Old World.

Inner tepals of fruiting perianth ± rounded, with a terminal
    spiny arista ; fruiting perianth 12–13 mm. broad in-
    cluding spines .   .   .   .   1. *E. australis*
Inner tepals of fruiting perianth lanceolate-muticous,
    without a terminal arista ; fruiting perianth not or
    scarcely exceeding 4 mm. broad including the spines .   2. *E. spinosus*

1. **E. australis** *Steinh.* in Ann. Sc. Nat., sér. 2, 9 : 195 (1838) ; Fl. Cap. 5 (1) : 481 (1912). Type : South Africa, Cape of Good Hope (collector & location uncertain)

An erect or diffuse, much branched, green herb, glabrous throughout. Stem furrowed. Ocreae 5 mm. long, membranous, brown, readily tearing and falling away. Leaves long-petiolate, ovate, with a rounded or obtuse apex, basally rounded, truncate or cordate, shortly decurrent to the petiole ; when mature with rounded basal lobes containing a broad basal sinus. Inflorescence rather laxly racemose, the flowers borne in axillary and some-times shortly pedunculate clusters (or terminally as a spiciform raceme if the upper leaves are absent). Male tepals green, in two subequal series, ovate-elliptic, 2 mm. long, scarcely exceeding 1 mm. in width, all with brown midrib and veins. Anthers ovate, brownish-orange ; filaments filiform, free. Female flowers sessile, all in axillary clusters. Inner 3 tepals erect, rounded or very broadly ovate, strongly veined with lateral veins bifurcating near the margin, the midrib excurrent as a short, rigid, spiny arista. Outer 3 tepals forming at the tips very sharp, rigid spreading spines, 4–5 mm. long ; the faces perforated at maturity with usually 4 large pits. Fruiting perianth parallel-sided, very accrescent and hard, 6 mm. long, 12–13 mm. overall broad. Fig. 1/1 and 2.

Kenya. Uasin Gishu District : Soy, *Brodhurst-Hill* 423 ! ; Kiambu District : Kabete railway station, Aug. 1947, *Bogdan* 1056 ! and near Kabete, alongside the Nairobi–Naivasha road, Nov. 1950, *Bogdan* 2846 !
Distr. **K**3, 4 ; introduced into our area, probably from South Africa, where it is reputedly native and known as Cape Spinach. Introduced into New Zealand, Mada-gascar, Australia (intentionally), and occasionally elsewhere as in Florida, California and Hawaii
Hab. By roads and railways ; essentially a plant of waste land ; 1800–1900 m.

Note. The leaves can be used as a vegetable, but the spiny perianths cause painful damage to cattle ; for this reason this species and the next should be prevented as far as possible from spreading.

2. **E. spinosus** (*L.*) *Campd.*, Monogr. Rum. 58 (1819) ; Post, Fl. Syr. Pal. Sin. : 696 (1896) ; Borg, Fl. Maltese Isl. : 115 (1927). Type : Crete, cult. Uppsala (LINN, lecto. !)

Very like the preceding in its ♂ flowers and vegetative characters, but markedly different in the fruiting perianth which is smaller, 5 mm. long and 4 mm. broad overall, narrowing upwards from a protruding rim shortly above the base. Inner tepals erect, lanceolate-muticous, without a terminal spiny arista. Outer tepals deeply perforated with usually 6 small pits, in 2 longitudinal rows, above the rim. Spines 1·5–2 mm. long, somewhat arcuately curved at the base and remaining a little reflexed or becoming horizontal. Fig. 1/3, p. 2.

KENYA. Nairobi, 1952, *Bally* 8216 !
DISTR. **K4** ; a weed of Mediterranean coastal countries, like the preceding introduced into our area.
HAB. Waste ground, etc. ; in general as for *E. australis* ; ± 1560 m.

SYN. *Rumex spinosus* L., Sp. Pl. : 337 (1753)

2. HARPAGOCARPUS

Hutch. & Dandy in K.B. 1926 : 363 (1926)

Hermaphrodite. Tepals 5, unequal, somewhat accrescent. Stamens 8. Ovary trigonous, with minute globular protuberances arranged in two rows down the angles ; these later developing into two rows of setae, with slightly deflexed, stellate, terminal barbs. Styles 3, subcapitate, later reflexing.

A monotypic genus confined to tropical Africa, so far known only from scattered localities.

H. snowdenii *Hutch. & Dandy,* in K.B. 1926 : 363 (1926) ; F.C.B. 1 : 423 (1948) ; F.W.T.A., ed. 2, 1 : 140 (1954). Type : Uganda, Elgon, Nkokonjeru, *Snowden* 946 (K, holo. !, BM, iso. !)

A perennial, rather weak, branched, climbing herb, nearly glabrous throughout. Stems greenish-brown. Ocreae brown, papery, obliquely truncate, 4–6 (–8) mm. long. Leaves sagittate, the lower ones broadly ovate-triangular, up to 9·3 × 5 cm., acutely acuminate, the upper more lanceolate, sometimes narrowly so, up to 15 × 3·3 cm. ; glabrous above, paler green and shortly and sparsely setose on the midrib and veins below. Petioles slender, variable in length, sometimes considerably exceeding the lamina. Inflorescence slender, branched, racemose, axillary and terminal ; the flowers borne singly or in pairs, erect-spreading at first, later pendant. Bracts 1·5–2 mm. long, membranous, truncate, 0·5–2 cm. apart. Pedicels filiform, 2–3 mm. long. Tepals 5, yellowish, unequal, connate towards the base, 2·5–3 5 × 2–2·5 mm., obovate, terminally rounded, later ± appressed to the fruit. Anthers orange, ± rounded ; filaments filiform, 1 mm. long. Ovary ovoid-ellipsoid, trigonous. Ripe fruit 6–8 × 4 mm., light orange-brown, shining, with two rows of purple setae up to 2 mm. long down the angles, each seta terminating in 2–4 reflexed, stellate barbs of unequal length, not exceeding 0·4 mm. long. Fig. 1/4–6, p. 2.

UGANDA. Toro District : Musandama [Msandama], Dec. 1925, *Maitland* 1013 ! ; Ruwenzori, Aug. 1938, *Purseglove* 205 ! and *Scott Elliott* 7855 ! and Jan. 1933, *Synge* 1611 ! ; Mbale District : Mt. Nkokonjeru, Dec. 1924, *Snowden* 946 !
KENYA. Meru, Dec. 1934, *Gedye in C.M.* 6700 !
TANGANYIKA. Morogoro District : Uluguru Mts., Morningside, near Morogoro, Dec. 1934, *E. M. Bruce* 356 !
DISTR. **U**2, 3 ; **K**4 ; **T**6 ; also in the Cameroons and Belgian Congo
HAB. Among undergrowth and shrubs in or at the edge of upland rain-forest ; 1350–2400 m.

SYN. *Fagopyrum ciliatum* Jac.-Fél. in Bull. Mus. Hist. Nat. 18 : 409, fig. 1–7 (1946). Type : French Cameroons, Bambuto Mt., *Jacques-Félix* 2692 (P, holo.)

3. RUMEX

L., Sp. Pl. : 333 (1753) & Gen. Pl., ed. 5 : 156 (1754)

Hermaphrodite or dioecious, usually glabrous herbs, often stout. Bracteoles 1 or 0. Tepals 6, in two series of 3 ; the outer small, non-accrescent ; the inner accrescent, erect, circular, ovate or triangular with the margin entire, wavy or toothed, sometimes winged, and the midrib sometimes swollen to produce a wart-like tubercle. Stamens 6, inserted at the base of the perianth. Nuts trigonous, enclosed within the inner tepals. Styles 3 ; stigmas penicillate or fimbriate.

A world-wide genus, most abundant in temperate regions.

Flowers dioecious, anemophilous, very small ; inflorescence a much branched, slender panicle ; low, decumbent or erect herb ; leaves small, largely basal .   .   .   .   .   . 1. *R. acetosella*
Flowers hermaphrodite :
  Leaves hastate or sagittate ; inflorescence a much branched panicle ; mature inner tepals ± pellucid, reddish :
    Leaves markedly trinervate, basal lobes small ; a shrubby herb, sometimes climbing .   . 2. *R. usambarensis*
    Leaves palmately nerved, basal lobes large ; a very stout perennial herb .   .   . 3. *R. abyssinicus*
  Leaves neither hastate nor sagittate ; inflorescence a simple or branched, stout panicle ; mature inner tepals opaque, dark brown :
    Inner tepals fringed with long teeth :
      Teeth terminally hooked :
        Leaves (5–) 7–9 times as long as broad ; inner tepals 3–4 mm. long ; nut 2–2·5 (–3) mm. long .   .   .   .   . 4. *R. bequaertii*
        (Leaves 3–5 times as long as broad ; inner tepals 5 mm. long ; nut 3 mm. long . *R. steudelii*) *
      Teeth not terminally hooked, but straight or slightly arcuate .   .   .   . 5. *R. dentatus*
    Inner tepals entire, or repand, or with a few shallow dentations, but without long teeth :
      Inner tepals entire, without tubercles .   . 6. *R. ruwenzoriensis*
      Inner tepals entire, repand or slightly dentate, with tubercles .   .   .   . 7. *R. crispus*

1. **R. acetosella** *L.*, Sp. Pl. : 338 (1753) ; C. H. Wright in Fl. Cap. 5 (1) : 475 (1912). Type : from Europe

A slender perennial, with erect or decumbent stems, seldom exceeding 30 cm. tall, arising from a basal tuft of leaves. Ocreae hyaline, 4–5 mm. long, readily tearing. Leaves long-petioled, hastate, often narrowly so, apically subacute, basally decurrent ; the terminal lobe seldom exceeding 4 cm. in length ; basal lobes much shorter, narrow, spreading outward and often turning forward ; immature leaves often unlobed, linear. Inflorescence slender, paniculate, much-branched, leafless ; the flowers borne in close fascicles up to 5 mm. apart. Male flowers 2–2·5 mm. across, the outer tepals oblong-lanceolate, partly appressed to the inner ; inner tepals greenish-brown becoming tinged with red. Anthers yellow, 1 mm. long,

* The occurrence of this species in tropical East Africa is open to doubt. See under *R. bequaertii*, p. 9.

FIG. 2. *RUMEX USAMBARENSIS*—**1,** flowering branch, × 2/3 ; **2,** fruit, showing tubercle at base of inner perianth, × 5 ; *R. ABYSSINICUS*—**3,** flowering branch, × 2/3 ; **4,** fruit, × 5 ; **5** and **6,** two different forms of leaf shape, × 2/3.

longer than the filaments. Inner tepals of ♀ flowers closely appressed to and accrescent with the nut ; outer tepals small, lanceolate. Nut light brown, sharply trigonous, ± 1 mm. long and approximately as broad.

KENYA. Nakuru District : Molo, July 1951, *Bogdan* 3177 (♀) ! and Upper Molo, July 1952, *G. R. C. van Someren in E.A.H.* 10913
DISTR. **K3** ; Europe and temperate Asia, North and South Africa, temperate America ; introduced into our area.
HAB. A weed in experimental grass plots ; 2400 m.

NOTE. *R. acetosella* (agg.) has been divided into four (? micro-) species differing, according to Löve in Hereditas 30 : 3 (1940), cytologically, geographically and morphologically. The above specimens have been identified as *R. angiocarpus* Murb. to which segregate most African material is apparently referable (Rech. f., Monogr. Rum. Afr. : 9 (1954)). It is not, however, decisively clear whether the subdivisions of *R. acetosella* are interdifferentiable in terms of morphology ; for this reason the Kenyan records are offered here in the aggregate sense.

2. **R. usambarensis** (*Dammer*) *Dammer* in E.J. 38 : 61 (1905) ; R. A. Graham in K.B. 1956 : 254 (1956). Type : W. Usambara Mts., Mlalo, *Holst* 2429 (K, lecto. !)

A shrub or straggling glabrous climber, up to 3 m. tall or more. Stems brown. Leaves astringent, petiolate, often clustered, narrowly to broadly elliptic with a hastate base, a little narrowed above the basal lobes, up to 5 (-9) cm. long, apically acute ; lobes reflexed, small, not exceeding 5 × 2 mm. ; leaves markedly trinervate (except in very narrow leaves), the lateral nerves arising arcuately from the base of the midrib and remaining complementarily parallel throughout most of their length. Petioles 1–4 cm. long. Inflorescence a much-branched, ± slender, leafless panicle. Flowers in fascicles, on filiform pedicels up to 5 mm. long. Outer tepals 1·75–2 mm. long, ovate, obtuse, later reflexing. Inner tepals wing-like, 5–7 mm. in diameter, ± pellucid, subequal, circular when mature, with a squarish basal sinus 1–1·5 mm. deep, red or reddish-brown, reticulately veined and with a small wart-like reflexed protuberance at the base. Nut trigonous, ovoid, 2–2·5 × 1 mm., brown, shining. Fig. 2/1 and 2.

UGANDA. Kigezi District : Kisoro, June 1939, *Purseglove* 746 ! ; Teso District : Serere, May 1932, *Chandler* 698 ! ; Mengo District : Entebbe, July 1922, *Maitland* 4 !
KENYA. Nakuru District : around Lake Elmenteita, Aug. 1947, *Bogdan* 988 ! ; Machakos District : Kima, Mar. 1930, *Napier* 32 ! ; Masai District : near Murueshi, Nov. 1932, *C. G. Rogers* 79 !
TANGANYIKA. Moshi District : Marangu, Mar. 1943, *Greenway* 3890 ! ; Lushoto District : Lushoto—Soni road near Nyassa bridge, June 1953, *Drummond & Hemsley* 3010 !; Mbeya, Mar. 1932, *R. M. Davies* 428 !
DISTR. **U2–4** ; **K3–4, 6** ; **T2–4, 7** ; also in Nyasaland & Belgian Congo
HAB. Open mist-forest, also upland grassland, bushland and exposed rocky slopes, 870–2400 m. ; occasionally cultivated for its medical properties.

SYN. [*R. maderensis* sensu F.T.A. 6 (1) : 115 (1909) ; F.D.O.-A. 2 : 198 (1932) F.P.N.A. 1 : 113 (1948); *non* Lowe (1838)]
     R. nervosus Vahl var. *usambarensis* Dammer in P.O.A. C : 169 (1895)
     R. trinervius Rech. f. in Oest. Bot. Zeitschr. 99 : 523 (1952) & in Bot. Not., Suppl. 3 (3) : 17 (1954). Type : Tanganyika, Ufipa District, Mbisi Mts., *Michelmore* 711 (K, holo. !)

NOTE. This species has been confused with *R. maderensis* Lowe, an endemic of Madeira and the Canary Islands, from which it is immediately told by the presence of the small protuberance at the base of the inner tepals, and by the marked trinervation of the leaves.

3. **R. abyssinicus** *Jacq.*, Hort. Vindob. 3 : 48, tab. 93 (1776) ; F.T.A. 6 (1) : 114 (1909) ; F.P.N.A. 1 : 112 (1948) ; F.C.B. 1 : 398 (1948) ; F.W.T.A., ed. 2, 1 : 139 (1954) ; Rech. f. in Bot. Not., Suppl. 3 (3) : 29 (1954). Type : a cultivated plant from Ethiopia (W, holo. †)

A large, very stout, perennial herb, up to 4 m. tall. Stems glabrous, green or reddish-green, up to 3 cm. wide at the base. Leaves petiolate, large, up to ± 30 × 20 cm., lush, glabrous or papillose, usually triangular-hastate but varying to sagittate, scutate, sublinear, or more rarely ovate ; the lobes spreading, more rarely directed forward ; apically acute to very obtuse ; with a variably deep and wide basal sinus ; primary nerves palmately arranged. Petioles long, up to 14 cm., those of the lower leaves often exceeding the lamina. Inflorescence a large, much branched, leafless panicle, oblong or pyramidal in outline, up to 40 cm. long, 25 cm. or more broad. Flowers borne in fascicles, on filiform pedicels up to 5 mm. long. Outer tepals ± ovate, 1·5 mm. long, membranous and brown, later reflexing. Inner tepals wing-like, accrescent, elongate-rounded, ± pellucid, reticulately veined, green becoming reddish-brown or brown, 5·5 × 4·5 mm. (7 × 7 mm., *fide* Rech. f.), with a basal sinus up to 1 mm. deep, and bearing a small reflexed protuberance at the base. Nut acutely trigonous, 2·25–3 × 2 mm., shining, light brown or dark with lighter angles. Fig. 2/3–6, p. 6.

Uganda. West Nile District : Lendu, Zeio [Zeu], Apr. 1940, *Eggeling* 3910 ! ; Kigezi District : Mt. Mgahinga, June 1949, *Purseglove* 2936 ! ; Mengo District : Bukalasa, Oct. 1931, *Hancock* 2329 !
Kenya. Nakuru District : Molo, Jan. 1912, *G. S. Baker* 367 ! ; North Nyeri District : Nyeri, Dec. 1921, *Fries* 185 ! ; Machakos District : Chyulu Hills, May 1938, *Bally* 7905 !
Tanganyika. Moshi District : Mt. Kilimanjaro, Bismarck Hut to Marangu, March 1934, *Greenway* 3893 ! ; Lushoto District : Kwamshemshi–Sakare road, July 1953, *Drummond & Hemsley* 3185 ! ; Morogoro District : Uluguru Mts., Lukwangule Plateau, Jan. 1934, *Michelmore* 913 !
Distr. U1, 2, 4 ; K3, 4, 6 ; T1–4, 6–8 ; widely spread in the highlands of tropical Africa, also in Madagascar
Hab. Upland grassland, margins of upland rain-forest & upland secondary bushland ; 750–3300 m.

Syn.　*R. schimperi* Meisn. in DC., Prodr. 14 : 67 (1856). Type : Ethiopia, Semen, Debra Eski, *Schimper* 514 (K, syn. !)
　　　R. abyssinicus Jacq. var. *schimperi* (Meisn.) Asch. in Schweinf., Beitr. Fl. Aethiop. : 171 (1867)
　　　R. abyssinicus Jacq. var. *angustisectus* Engl., Hochgebirgsfl. Trop. Afr.: 203 (1892); F.P.N.A. 1 : 112 (1948). Type : Ethiopia, Bagemdir, Debra Tabor, *Schimper* 1527 (K, iso. !)
　　　R. abyssinicus Jacq. var. *mannii* Engl., l.c. 203. Type : Cameroon Mt., *Mann* 1217 (K, iso. !)
　　　R. abyssinicus Jacq. var. *kilimandschari* Engl., l.c. 203. Type : Tanganyika, Kilimanjaro, *Meyer* 300 (B, holo.)
　　　? *R. hastatus* Peter, F.D.O.-A. 2 : 196 (1932), *non* Don (1825), *nom. illegit.* Type : Tanganyika, Buha District : between Bikare & Mkigo, *Peter* 38768 (B, †)
　　　R. abyssinicus Jacq. var. *calystegiifolius* Rech. f. in Bot. Not., Suppl. 3 (3) : 35 (1954). Type : Eritrea, Serae, *Pappi* 237 (FI)
　　　R. abyssinicus Jacq. var. *retrorsilobatus* Rech. f. in Bot. Not., Suppl. 3 (3) : 36 (1954). Type : Ethiopia, Dembia, *Chiovenda* 1886 (FI)

Note. The considerable variation in leaf shape has prompted the creation of varieties. But these seem merely to point stages in the general polymorphy of the leaf shape alone, and as there are apparently no other characters correlatable to those of leaf shape, it is proposed to sink the varieties, as listed above, into synonymy with the species.

4. **R. bequaertii** *De Wild.*, Pl. Bequaert. 5 : 2 (1929) ; F.P.N.A. 1 : 116 (1948) ; F.C.B. 1 : 400 (1948) ; Rech. f. in Bot. Not., Suppl. 3 (3) : 93 (1954) ; F.W.T.A., ed. 2, 1 : 139 (1954). Type : Belgian Congo, Mukule, *Bequaert* 5905 (BR, lecto., *fide* Rech. f.)

An erect, ± glabrous, stout, perennial herb, up to 1·8 m. tall. Stems green to greenish brown. Leaves narrow and long, up to 33 cm. long but rarely exceeding 6 cm. broad (usually approximately 7–9 times as long as broad). oblong-lanceolate, often ± parallel sided, apically obtuse, sometimes rounded, basally cuneate ; glabrous or with scattered papillae on the under-surface ; flat or crispate on the margins. Inflorescence an open, ± stout panicle with long branches. Flowers in fascicles, pendulous. Inner tepals accrescent, dark brown, opaque, elongate-triangular, 3–4 (–5) mm. long, with or without tubercles, usually with an unarmed, obtuse, lanceolate apex but the margins otherwise armed with 5–6 strongly hooked teeth 1·5–2 mm. long. Nut trigonous, ovoid, brown, shining, 2·5–3 × 1·5–2 mm.

UGANDA. Toro District : Bwamba, Oct. 1925, *Fyffe* 37 ! ; Mbale District : Bulam-buli, Sept. 1932, *A. S. Thomas* 566 ! ; Masaka District : Malabigambo Forest, near Katera, Oct. 1953, *Drummond & Hemsley* 4552 !
KENYA. Nakuru District : near Thomson's Falls, Oct. 1931, *Pierce* 1462 ! ; Kiambu District : Limuru, Feb. 1915, *Dummer* 1603 ! ; Kisumu-Londiani District : Lumbwa, Bondui, Mau Forest, Jan. 1946, *Bally* 4992 !
TANGANYIKA. Masai District : Ngorongoro crater, Apr. 1941, *Bally* 2302 ! ; Lushoto District : Mkuzi, Apr. 1953, *Drummond & Hemsley* 2169 ! ; Rungwe District : Mbeye, below Poroto Mts., Mar. 1932, *St. Clair-Thompson* 820 !
DISTR. U2–4 ; K3–5, ? 6 ; T2, 3, 6–8 ; widely spread through eastern Africa reaching Ethiopia and the Transvaal ; also in the Cameroons and Madagascar
HAB. Upland grassland & bushland, upland rain-forest, riverside grassland, particularly in damp places, 690–3700 m.

SYN. [*R. nepalensis* sensu F.T.A. 6 (1) : 117 (1909), *non* Spreng.]
    R. quarrei De Wild., Pl. Bequaert. 5 : 3 (1929). Type : Belgian Congo, Katanga, Kufubu, *Quarré* 269 (BR, lecto. *fide* Rech. f.)
    R. camptodon Rech. f. in Bot. Centr., Beih. 49 (2) : 76 (1932). Type : Mt. Kenya, near Forest Station, *Fries* 583 (S, syn.)
    R. bequaertii De Wild. var. *quarrei* (De Wild.) Robyns in F.C.B. 1 : 401 (1948) ; F.P.N.A. 1 : 117 (1948)

NOTE. *R. quarrei* has been distinguished from *R. bequaertii* by its having large tubercles on all the inner tepals (rarely only on one), whereas in typical *R. bequaertii* they are absent or, as it seems, only in a most rudimentary state of development. *R. camptodon* lies between the two in having a large tubercle on one tepal and smaller ones on the others. Apparently the matter is simply one of the degree of accrescence of the midrib, and when material is very immature it is sometimes impossible to forecast which " species " is being handled. It is therefore proposed to sink *R. camptodon* and *R. quarrei* into synonymy with *R. bequaertii*. The tubercled form is perhaps commoner than that wholly without tubercles.
    Certain examples from Kilimanjaro (*Volkens* 676 ! ; *Greenway* 3860 !) have broader leaves, inner tepals 4–5 mm. long, and nuts 2·75–3 mm. long, thus suggesting an approach to *R. steudelii* Hochst. ex A. Rich. Both have however been determined as *R. bequaertii* by Rechinger, apart from which the presence in our area of *R. steudelii* —a dock of Ethiopia, Somaliland, and South Africa—as an isolated occurrence would be rather unlikely. Further collecting on Kilimanjaro, especially in the altitude range 2400–2700 m., is nevertheless desirable. Another example, from Kenya, neighbourhood of Thomson's Falls, *Blain in E.A.H.* 10915, has basal leaves 9 cm. broad, inner tepals 6–5 mm. long, and nuts 2·5–3 mm. long, and has much the appearance of *R. steudelii*. This example may merely represent a large form of *R. bequartii*, to which, following Rechinger, it is here referred.

5. R. dentatus *L.*, Mant. 2 : 226 (1771) ; Campd., Monogr. Rum. : 81 (1819); Rech. f. in Bot. Centr., Beih. 49 (2) : 12 (1932) and in Bot. Not., Suppl. 3 (3) : 100 (1954). Type : Egypt, cult. Uppsala (LINN !)

A stout annual (? also biennial), glabrous herb, 20–70 cm. tall. Stems often branched. Basal leaves ovate-oblong, often rather narrow, up to 18 × 4 cm., flat or crisped along the edges, apically obtuse to subacute, basally truncate or more rarely subcordate ; cauline leaves becoming progressively narrower and shorter, the uppermost sublinear. Petioles long, but usually shorter than the lamina. Inflorescence an open, rather stout panicle, leafy or not. Inner tepals varying from ovate-triangular to deltoid, 3·5–5 × 2–3 mm., equally or unequally tubercled, the margin entire or more usually armed with straight or slightly arcuate teeth (occasionally slightly hooked) which in length are less than or exceed the width of the tepal. Nut 2–2·75 × 1·5 mm., brown, acutely trigonous.

subsp. **dentatus**

Inner tepals 3–4 × 2 mm., acute, all with tubercles nearly covering the lamina ; the teeth shorter than the width of the tepal.

UGANDA.   Mengo District : Kivuvu, Nov. 1914, *Dummer* 1231 !
DISTR.   U4 ; Egypt, Libya and Palestine.   Presumably introduced in Uganda
HAB.   River banks, oases, generally damp places.   The Uganda gathering was from a " lowland, grassy swamp ", at 1200 m.

SYN.   *R. callosissimus* Meisn. in DC., Prodr. 14 : 57 (1854).   Type : Egypt, *Ehrenberg* (G–DC, syn.)
       R. dentatus L. subsp. *callosissimus* (Meisn.) Rech. f. in Bot. Centr., Beih. 49 (2) : 13 (1932)

NOTE.   Unfortunately the single specimen is a poor one, but it appears to be this sub-species.   Further material is desired.

6. **R. ruwenzoriensis** *Chiov.* in Bull. Soc. Bot. It. 1917 : 56 (1917) ; Rech. f. in Bot. Not., Suppl. 3 (3) : 59 (1954).   Type : Uganda, Ruwenzori, Mubuku Valley, *Savoia* (TO, syn.)

A stout, erect herb, 1 m. or more tall, with flowering branches arising from the axils of the lower leaves.   Stem greenish brown to brown, glabrous. Lower leaves large, on petioles up to about 20 cm. long, ovate-oblong, up to 36 × 12 cm., apically acute or acuminate, basally rounded, truncate or subcordate, glabrous except for scattered papillae on the nerves of the sub-surface ; cauline leaves on shorter petioles, ovate-lanceolate to elliptic-lanceolate, progressively smaller, the uppermost ± linear and nearly sessile. Inflorescence a stout, usually rather dense, branched panicle, often leafy below and sometimes throughout.   Inner tepals ovate-lanceolate, ± obtuse, 4–5 (–7 *fide* Chiov.) mm. long, without tubercles, the margin entire, un-armed.   Nut 2 × 1 mm., acutely trigonous, dark brown, shining.

UGANDA.   Toro District : Ruwenzori, Bujuku Valley, May 1933, *Eggeling* 1295 !, 1296 ! ; Mbale District : Elgon, Bulambuli, Aug. 1929, *Saundy & Hancock* 58 ! and Aug. 1934, *Synge* 1010 !
KENYA.   North Nyeri District : Mt. Kenya, Naro Moru, Sept. 1948, *J. G. Williams in Bally* 6396 ! ; Aberdare Mts., *James* ! ; Mt. Kenya, Dec. 1943, *Bally* 3359 !
TANGANYIKA.   Morogoro District : Lukwangule Plateau, Mar. 1953, *Drummond & Hemsley* 1508 ! & Mar. 1934, *Michelmore* 898 ! ; Mbeya District : Kiwara River, Lower Fishing Camp, Oct. 1947, *Greenway & Brenan* 8265 !
DISTR.   U2, 3 ; K3, 4 ; T6, 7 ; also in mountains of the Belgian Congo
HAB.   Upland grassland & upland moor, bamboo-forest, preferring moist situations ; 1950–3700 m.

SYN.   *R. afromontanus* T. C. E. Fries in N.B.G.B. 9 : 35 (1924) ; F.P.N.A. 1 : 115 (1948) ; F.C.B. 1 : 400 (1948).   Type : Kenya, Mt. Kenya, between Coles Mill & Forest Station, *Fries* 651a (K, syn. !)

7. **R. crispus** *L.*, Sp. Pl. : 335 (1753) ; Lousley in B.E.C. Rep. 1941–42 : 552 (1944) ; Rech. f. in Bot. Not., Suppl. 3 (3) : 76 (1954).   Type : un-certain, Europe, possibly from Sweden

A stout, ± glabrous perennial.   Stems brownish-red to greenish-brown. Leaves sometimes tending to redden, glabrous except for papillae on the veins of the subsurface, narrowly oblong-lanceolate, 15–26 × 2–5·5 cm. (–35 × 8 *fide* Rech. f.), the upper ones often linear ; crisped on the margins ; apically ± acute, basally attenuate, cuneate, or rounded, often rather unequally so.   Lower petioles up to 20 cm. long ; those of the stem leaves much shorter, often 1·5–3 cm.   Inflorescence a rather dense, branched panicle.   Inner tepals 3·5–5 mm. long, and as broad or nearly so, ovate-deltoid, usually all tubercled, the margins entire, undulating, or very slightly and shallowly toothed towards the base.   Nut 2·5–3 × 1·5–2 mm., medium brown, acutely trigonous.

KENYA. Nairobi, Aug. 1945, *Bally* 4621 !
DISTR. K4 ; a native of Europe and western Asia ; elsewhere widely introduced.
HAB. A plant of damp places, and a weed of disturbed ground ; at Nairobi it occurs
" in and along the riverbed," at 1620 m.

NOTE. One of Bally's examples has been determined by Rechinger f. as " *Rumex*
probably *crispus* × ?". Further gatherings are desired.

4. POLYGONUM

L., Sp. Pl. : 359 (1753) & Gen. Pl., ed. 5 : 170 (1754)

Hermaphrodite herbs or (rarely) shrubs, glabrous to very hairy, slender
or stout, often paludal. Ocreae with or without a terminal fringe of stiff
bristles. Flowers (in. E. African spp.) borne in spiciform racemes or terminal
capitula. Bracteoles 2, conjoined, thinly membranous and contained
within the ocreiform bracts. Perianth (in E. African spp.) not accrescent,
calyciform at the base, with 4–5, often imbricate, petaloid, ± persistent
tepals. Stamens 5–8, included or protruding, a little dilated at the base or
conjoined there into a ring. Styles 2 or 3, often united for part of their
length ; stigmas capitate. Ripe nuts often black and shining, acutely to
obscurely trigonous, or lenticular with convex or concave (dimpled) sides.

A multispecific genus occurring throughout the greater part of the world.

Herbs or low bushes, procumbent, decumbent, erect,
  or growing in water with floating leaves, sometimes climbing but not twining :
Flowers white, in small terminal capitula ; each
  capitulum usually subtended by an involucral
  leaf ; petioles winged .   .   .   .   1. *P. nepalense*
Flowers usually pink or red, borne in spiciform
  racemes ; if short and ± capitate, then
  without an involucral leaf :
Bracts all foliferous :
  Stems woody ; ocreae brown throughout ;
    prostrate or climbing (not twining) shrub   2. *P. afromontanum*
  Stems herbaceous ; ocreae silvery ; procum-
    bent or decumbent herbs :
    Nut 1·5–2 mm. long, black, shining .   .   3. *P. plebeium*
    Nut 3–3·5 mm. long, dark brown, matt   .   4. *P. aviculare*
Bracts not foliferous :
  Peduncles with stalked glands :
    Racemes short, subcapitate, dense ; in-
      florescence dichotomously branched ;
      leaves basally attenuate, glabrous   .   5. *P. glomeratum*
    Racemes very slender, spiciform, the bracts
      not contiguous, sometimes ± capitate
      terminally ; leaves usually truncate
      to sagittate at the base, and rough
      with bristly hairs   .   .   .   .   6. *P. strigosum*
  Peduncles without stalked glands :
    Racemes slender, zigzag when immature,
      2–5 together ; leaves long and narrow,
      linear-lanceolate to linear-elliptic,
      basally attenuate .   .   .   .   7. *P. salicifolium*

Racemes dense, stout or ± so, elongated or
     short ; leaves ovate, ovate-elliptic,
     elliptic, lanceolate or oblong-lanceolate,
     basally attenuate or rounded :
Nut lenticular with dimpled faces ;
     plant stout ; leaves glabrous to
     thickly ashen or white-tomentose .   8. *P. senegalense*
Nut lenticular with convex faces, or
     trigonous or obscurely so ; leaves
     glabrous to thickly tomentose, but
     not white :
  Leaves basally rounded to cordate ;
       racemes short .   .   .   .   9. *P. amphibium*
  Leaves basally attenuate, never
       rounded :
    Ocreae with a leafy limb   .   . 10. *P. limbatum*
    Ocreae without a leafy limb :
      Racemes elongated, usually con-
           siderably exceeding 3 cm. in
           length :
        Peduncles and leaves hairy ;
             leaf subsurfaces with soft
             (or sometimes ± bristly)
             hairs on the midrib ;
             racemes elongated ; peri-
             anth 3–5 mm. long .   . 11. *P. pulchrum*
        Peduncles usually glabrous ; leaf
             subsurfaces with rather
             hard bristles on the mid-
             rib ; racemes elongated ;
             perianth usually not ex-
             ceeding 3 mm. in length . 12. *P. setosulum*
      Racemes short, 1–3 cm. long,
           often little more than capi-
           tate ; plant ± glabrous
           throughout   .   .   . 13. *P. persicaria*
Herbs, climbing and twining :
  Outer tepals broadly winged, the wings decurrent
       on the pedicel .   .   .   .   .   . 14. *P. baldschuanicum*
  Outer tepals keeled ; not or only slightly winged,
       and if so the wings not decurrent on the
       pedicel .   .   .   . 15. *P. convolvulus*

1. **P. nepalense** *Meisn.*, Monogr. Polygon. : 84 (1826) ; F.W.T.A., ed. 2, 1 :
140 (1954). Type : Nepal, *Wallich* (G–DC, holo.)

A rather slender, branched, somewhat straggling annual. Stems basally
decumbent and rooting at the nodes, pale green to greenish brown, glabrous
or with scattered gland-tipped hairs usually more numerous below the
ocreae. Ocreae glabrous, brown, membranous, without a terminal fringe of
setae, 6–8 (–10) mm. long. Leaves petiolate, ovate or ovate-deltoid, up to
5 × 3 cm., evenly but often abruptly narrowed to an acute apex, basally
truncate or abruptly narrowed to the petiole and with basal auricles. Petioles
winged, up to 1·5 cm. long and to 3·5 mm. wide. Inflorescence capitate, the
capitula 6–9 mm. broad and terminating the branches singly or in pairs,
each capitulum subtended by an involucral, sessile leaf. Peduncles with
deflexing gland-tipped hairs below the capitula, otherwise glabrous or

largely so. Bracts glabrous, ovate-lanceolate, narrowed to an acute apex, 5 mm. long, with a broad, hyaline margin. Flowers bluish-white or white, about 12 in each capitulum. Perianth ± 3 mm. long, urceolate, with 4 tepals free above the middle, the outer pair folded and the inner pair flat. Stamens included ; the anthers bluish-black, ± rounded ; filaments white, filiform. Styles 3, connate nearly to the apex. Nut dark-brown, lenticular, minutely areolate, 2 × 2 mm. Fig. 3/2, p. 15.

UGANDA. Toro District : Ruwenzori, Mubuku Valley, Jan. 1939, *Loveridge* 300 ! ; Ankole District : Igara, Mitoma, Oct. 1938, *Purseglove* 419 ! ; Masaka District : Sese Islands, Bukasa, June 1932, *A. S. Thomas* 139 !
KENYA. Nakuru District : E. Mau Forest Reserve, Sept. 1949, *Maas Geesteranus* 5934 !, 6119 ! ; Kiambu District : banks of Katamayu River, Limuru, Feb. 1948, *Bogdan* 1513 ! ; Kericho District : SW. Mau Forest Reserve, Aug. 1949, *Maas Geesteranus* 5630 !, 5726 !
TANGANYIKA. Moshi District : Lyamungu, Oct. 1943, *Wallace* 1102 ! ; Morogoro District : Uluguru Mts., Mgeta River in gorge below Hululu Falls, Mar. 1953, *Drummond & Hemsley* 1582 ! ; Rungwe District : below S. slopes of Poroto Mts., Mbeye, Mar. 1932, *St. Clair-Thompson* 813 !
DISTR. U2, 4 ; K3–5 ; T2, 6–8 ; throughout tropical Africa and tropical Asia, Madagascar, South Africa (? introduced)
HAB. By streams and rivers, in marshes, grasslands and at forest edges, also as a weed of cultivation ; 1140–2700 m.

SYN. *P. punctatum* Buch.-Ham. ex D. Don, Prodr. Fl. Nepal : 72 (1825), *non* Raf. (1820), var. *alatum* D. Don, Prodr. Fl. Nepal : 72 (1825), *nom. illegit.* Type : Nepal, *Buchanan-Hamilton* (location doubtful)
   P. perforatum Meisn., Monogr. Polygon. : 83 (1826)
   P. alatum Buch.-Ham. ex Spreng., Syst. Veg., cur. post. : 154 (1827) ; F.T.A. 6 (1): 104 (1909) ; F.P.N.A. 1 : 118 (1948), *nom. illegit.*

2. **P. afromontanum** *Greenway* in K.B. 1952 : 355 (1952). Type : Kenya, Aberdare Mts., *Fries* 1271 (UPS, holo., K, iso. !)

A low shrub with prostrate and climbing branches up to 1m. long. Stems red-brown above, brown and woody below, minutely puberulent. Ocreae strongly nerved, brown, membranous, readily tearing, ± 1 cm. long. Leaves small, nearly sessile, glabrous, all of ± similar size, rather leathery, 1–4 cm. × 5–10 mm., ovate-elliptic, ± evenly narrowed to each end ; apically acute (or obtuse when immature) with a sharp apical mucro ; margins revolute. Inflorescence a leafy raceme ; the flowers all axillary, 3–5 in each cluster, yellow-green or whitish, becoming brownish-purple. Pedicels ± 1 mm. long. Tepals 2·5–2·75 mm. long, ovate, very obtuse ; free for ¾ the length of the perianth. Stamens 6–8, dilated below. Styles 3, free to the base ; stigmas capitate. Nut trigonous, brown, smooth and shining ; at least two faces very concave. Fig. 3/5, p. 15.

KENYA. Mt. Kenya, Sept. 1943, *J. Bally in Bally* 3246 ! ; W. side Mt. Kenya, between Coles Mill and Forest Station, Jan. 1922, *Fries* 946 ! ; Aberdare Mts., *Heller* !
TANGANYIKA. Masai District : Oldeani Mt., Nov. 1956, *Greenway* 9071 ; Mbulu District : Mt. Hanang, below Werther's Peak, June 1948, *Greenway* 7719 !
DISTR. K?3, 4 ; T2 ; known only from the above-mentioned localities
HAB. Upland moor, upland bushland and at edges of rain-forest ; 2100–3150 m.

SYN. *P. paradoxum* T. C. E. Fries in N.B.G.B. 9 : 33 (1924), *non* Léveillé (1909), *nom. illegit.* Type : as *P. afromontanum* Greenway

3. **P. plebeium** *R. Br.*, Prodr. : 420 (1810) ; F.T.A. 6 (1) : 105 (1909) ; Hook. f., Fl. Br. Ind. 5 : 27 (1890) ; F.W.T.A., ed. 2, 1 : 140 (1954). Type : Australia, *R. Brown* 2994 (BM, holo. !)

A glabrous, much branched, prostrate, leafy, annual herb. Stems scabrid, reddish-brown. Ocreae silvery in the upper part or throughout most of their length, with concolorous veins, very readily lacerating, fringed with laciniae of varying length up to 2 mm. long. Leaves very small, up to 13 × 2·5 mm., all of a similar size, linear to narrowly obovate-elliptic, marginally revolute, glabrous and rather leathery, dark green becoming red ; midrib obvious but lateral veins scarcely perceptible. Inflorescence a shortly branched, leafy, congested raceme. Flowers in axillary clusters 1–3 (–5) together. Perianth greenish, 2 mm. long. Tepals 4, ± 1·5 mm. long, lanceo-late-elliptic, the outer pair keeled, the inner pair flat. Styles 3, free, ± 3 mm. long. Nut trigonous, smooth, black, shining, 1·5–2 mm. long. Fig. 3/6 and 7.

TANGANYIKA. Kondoa District : Bubu Valley near Salia, Dec. 1927, *B. D. Burtt* 828 ! ; N. of Lake Rukwa, Kisangu river-bed, *Richards* 3526 ! ; Morogoro District : Vikoza Forest Reserve, Mar. 1955, *Mgaza* 29 !
DISTR. T4–6; Nyasaland, Rhodesia and Madagascar; also in Afghanistan, India, Philippines and Australia
HAB. Apparently confined to dried river-beds and drying mud-flats beside lakes ; possibly introduced ; 690–1600 m.

4. **P. aviculare** *L.*, Sp. Pl. : 362 (1753) ; F.T.A. 6 (1) : 105 (1909) ; Fl. Cap. 5 (1) : 464 (1912). Type : from Europe

A glabrous, much branched, prostrate or decumbent-ascending annual. Stems smooth. Ocreae hyaline, silvery in the upper half, readily lacerating, the veins brownish throughout. Leaves not leathery ; narrowly elliptic, linear-oblong or linear, acute or obtuse, very variable in size, 1–5 × 0·2–1·7 cm., either all of a similar size or with the stem leaves larger than those of the flowering branches, sometimes marginally revolute, the lateral veins readily perceptible. Inflorescence a branched leafy raceme, the bracts congested or more often separated up to about 2 cm. apart. Perianth pink, white or greenish white, (2–) 3–4 mm. long. Tepals terminally rounded, ovate-oblong, ⅔ length of the perianth, white or pink with a green centre, the outer pair keeled, the inner pair flat. Styles 3, free, almost sessile. Nut 3–3·5 mm. long, trigonous, dark reddish-brown, matt. Fig. 3/8.

KENYA. Naivasha District : Aberdare Mts., Kinangop, Aug. 1947, *Bally* 5238 !
DISTR. K3 ; a native of temperate Europe, but now so widely introduced elsewhere as to have become cosmopolitan.
HAB. A weed of arable and disturbed land ; introduced into Kenya ; 2580 m.

NOTE. *P. aviculare* has been divided into a number of microspecies (see Lindm. in Sv. Bot. Tidskr. 673 *et seq.* (1912)). The above record can perhaps be referred to *P. aequale* Lindm., but this and the other segregates of the Linnean species are rather inconclusively differentiable.

5. **P. glomeratum** *Dammer* in F.R. 15 : 386 (1919) ; De Wild., Pl. Bequaert. 5 : 255 (1931) ; F.C.B. 1 : 413 (1948). Types : French Cameroons, *Ledermann*, several numbers, (B, syn.)

An erect herb, 1–1·5 m. tall. Stem reddish-brown, glabrous. Ocreae 2·5–3 cm. long, ovate-lanceolate, reddish-brown, subglabrous, with a terminal fringe of short setae up to 1 mm. long, readily tearing down one side, the mouth not closely appressed. Leaves petiolate, narrowly elliptic-lanceolate, 7·5–15 × 1–2·5 cm., narrowed to each end, apically acute, basally gradually or abruptly attenuate, glabrous or with a few scattered, short setae on the margins. Petioles glabrous, up to 2·5 cm. long, shorter upwards, the upper leaves subsessile. Inflorescence a dichotomously branched panicle of short, dense, racemose capitula, the peduncles short, slender, covered with stalked glands, the terminal peduncle scarcely exceeding 1 cm. Bracts obscurely truncate, 1·25–3 mm. long, with a shortly-ciliate terminal fringe. Perianth white or pink ; tepals 5, linear-lanceolate, 2·8–3 mm. long, longer than the tube. Stamens 5–7, included. Styles 2, connate above the middle ; stigmas capitate. Nut biconvex-lenticular, 1·5–1·75 mm. long. Fig. 3/3.

FIG. 3. **1,** *POLYGONUM SALICIFOLIUM,* × 2/3 ; **2,** *P. NEPALENSE,* × 2/3 ; **3,** *P. GLOMERATUM,* × 2/3 ; **4,** *P. STRIGOSUM,* × 2/3 ; **5,** *P. AFROMONTANUM,* × 2/3 ; **6,** *P. PLEBEIUM,* × 2/3 ; **7,** *P. PLEBEIUM,* fruit, × 3 ; **8,** *P. AVICULARE,* fruit, × 3.

TANGANYIKA.   Ulanga District : Masagati, June 1931, *Schlieben* 1134! ; Rungwe
District : Mwakete, Feb. 1911, *Stolz* 616! ; Songea District : about 5 km. E. of
Songea, Apr. 1956, *Milne-Redhead & Taylor* 9505 !
DISTR.   T6–8 ; also in Nyasaland, Belgian Congo, Nigeria, British and French Cameroons
and French Guinea
HAB.   Wet places, stream sides, sometimes in flowing water; 1000–2000 m.

SYN.   [*P. pedunculare* sensu F.T.A. 6 (1) : 107 (1909) excl. var. ; De Wild., Contrib.
Fl. Katanga, Suppl. 3 : 106 (1930) & Pl. Bequaert. 5 : 256 (1931); *non* Wall.
ex Meisn.]
[*P. strigosum* R. Br. var. *pedunculare* sensu F.W.T.A., ed. 2, 1 : 141 (1954),
non (Wall. ex Meisn.) Steward]

NOTE.   There is some resemblance between this species and the Asian *P. pedunculare*
Wall. ex Meisn., with which it has been confounded. The latter differs in having
broader leaves (8 × 2·8 cm.), eglandular (typically) or glandular peduncles, and in
the presence of retrorse bristles at the base of the ocreae. *Schlieben* 1134 is very
immature, but seems to be *P. glomeratum*.

6. **P. strigosum** *R. Br.*, Prodr. :   420 (1810) ; F.T.A. 6 (1) : 106 (1909) ;
F.C.B. 1 : 412 (1948). Type : Australia, *R. Brown* 2997 (K, syn. !, BM,
syn. !)

A rather slender, erect annual herb, up to ± 1·5 m. tall. Stems simple or
branched, yellowish to greenish-brown, usually with retrorse, spiny bristles
down the angles interspersed with short- and long-stalked glands ; rarely ±
glabrous. Ocreae foliferous at the base, 2–3 cm. long, truncate, with scattered
bristly hairs interspersed with shorter hairs.   Leaves petiolate, narrowly
linear-lanceolate to lanceolate (broader in Australian examples), 8–10 ×
0·8–1·6 (–2·5) cm., truncate to sagittate, the basal lobes 0·5–1·2 (–2) cm. long,
usually reflexed, sometimes reduced to very small, lateral teeth, rarely absent
and then the leaves ± elliptic and basally attenuate, apically long-acute ;
degree of hairiness variable, both surfaces usually with at least a few hairs,
but the undersurface typically rough with spiny, recurved bristles on the
midrib and veins.   Petioles short, 0·2–1 (–2) cm. long, usually bristly.
Inflorescence a very lax and slender, interrupted, ± dichotomously-
branched panicle, up to ± 4·5 cm. long, or aggregated terminally into short,
few-flowered capitula, the peduncles covered with horizontally-spreading
gland-tipped hairs. Bracts ciliate. Flowers 1–3 together. Pedicels up to
5 mm. long, exceeding the bracts, with glandular hairs. Perianth 2·5–3 mm.
long, glandular ; tepals 5, rose pink or white edged with pink, ± equalling
the tube, broadly oblong, terminally rounded. Stamens 5, included. Styles
2, united for $\frac{2}{3}$ their length. Nut biconvex-lenticular or bluntly trigonous,
2·25 × 1·25 mm., smooth, shining. Fig. 3/4, p. 15.

UGANDA.   Ankole District : Igura, Feb. 1939, *Purseglove* 589! ; Kigezi District :
Kamuganguzi swamp, July 1952, *Norman* 130! ; Masaka District : Katera, Sango
Bay, Katera, June 1935, *A. S. Thomas* 1312!
KENYA.   Nairobi District : Karura Forest, Karura river-bank, Mar. 1950, *Rayner*
270! ; Kiambu District : Theta papyrus-swamp, *Battiscombe* 1125! ; W. side of Mt.
Kenya, near Nitki River, Feb. 1922, *Fries* 1948 !
TANGANYIKA.   Bukoba, Aug. 1931, *Haarer* 2076! ; Lushoto District : W. Usambara
Mts., Mkuzi, Apr. 1953, *Drummond & Hemsley* 2153 ! & Kwa Mshusa, Aug. 1893,
Holst 9026 !
DISTR.   U2, 4 ; K4 ; T1, 3, 7, 8 ; widely spread in tropical Africa ; also in Madagascar,
South Africa, Asia and Australia
HAB.   Wet places, river banks, etc., often growing in water ; 1110–1920 m.

SYN.   [*P. pedunculare* Wall. ex Meisn. var. *angustissimum* sensu F.T.A. 6 (1): 107 (1909)
& sensu F.D.O.-A. 2 : 191 (1932) ; *non* Hook. f.]
?*P. strigosum* R. Br. var. *sanguineum* Peter in Abh. Ges. Wiss. Göttingen, n.f.,
13 (2) : 52 (1928). Type : Tanganyika, Pare Mts., Tona, *Peter* 41460 (B,
holo. †)

P. pedunculare Wall. ex Meisn. var. *subsagittatum* De Wild., Contrib. Fl. Katanga, Suppl. 3 : 107 (1930). Type : Belgian Congo, *Scaetta* 641 (BR, holo. !)
P. strigosum R. Br. forma *sanguineum* (Peter) Peter in F.D.O-A. 2 : 189 (1938).

VARIATION. There is a gradual variation from typical material with sagittate leaves, recurved bristles on the stems and leaf subsurfaces, and glandular peduncles and pedicels, to forms such as *Eggeling* 868 ! which have basally attenuate leaves, are almost devoid of bristles, and have almost eglandular inflorescences. Intermediates are *Chandler* 1663 ! which has sagittate leaves, eglandular inflorescences, and is almost without bristles, and *Drummond & Hemsley* 2153 ! which although glandular has no bristles on the leaves which bear only rudimentary lobes at the base.

7. P. salicifolium *Willd.*, Enum. Hort. Berol. : 428 (1809) ; F.P.N.A. 1 : 119 (1948) ; F.C.B. 1 : 413 (1948) ; F.W.T.A., ed. 2, 1 : 141 (1954). Type : from Canary Islands, Tenerife, *fide* Willdenow

An erect or basally decumbent, slender annual, up to 1 m. tall. Stems green becoming brown below, simple or branched, glabrous. Ocreae appressed when young, later more open, up to 2 cm. long, membranous, brown, rather thinly covered with closely ascending, bristly hairs 1–1·25 mm. long, and terminally fringed with erect-patent, stiff bristles 1–2·5 cm. long. Leaves sessile or nearly so, narrowly lanceolate-elliptic, 8–15 × 1–2 cm. ; apically acute, basally attenuate, usually glabrous but with ciliate margins and veins on the undersurface, more rarely papillose or pubescent. Inflorescence a slender, spiciform raceme, 2–9 cm. long or more, the racemes often 2–5 together to appear somewhat digitate, little or not interrupted ; the angle of the bracts giving a markedly zigzag appearance to the racemes when immature. Bracts glabrous, truncate or truncate-rounded, reddish-brown, with a terminal fringe of rigid bristles ± 1 mm. long. Perianth rose-pink or white, 2–2·5 (–3) mm. long. Tepals 5, longer than the tube, broadly ovate. Stamens included. Styles 3, united for half their length ; stigmas orange. Nut trigonous or lenticular, smooth, shining, 2–2·5 × 1–2·5 mm. Fig. 3/1, p. 15.

UGANDA. Kigezi District : Kashambya, July 1949, *Purseglove* 2990 ! ; Toro District : Nyakasura School, June 1932, *Shillitoe* 113 ! ; Mengo District : Entebbe, Sept. 1922, *Maitland* 193 !
KENYA. Kiambu District : Kabete, *Mettam* 261 ! ; Embu District : Emberre, Sept. 1932, *M. D. Graham* 2160 ! ; Kisumu–Londiani District : Tinderet Forest Reserve, June 1949, *Maas Geesteranus* 5105 !
TANGANYIKA. Shinyanga, Nov. 1938, *Koritschoner* 2049 ! ; Buha District : Kaberi swamp, Aug. 1950, *Bullock* 3127A ! ; Morogoro District : Uluguru Mts., Mgeta River valley, above Bunduki, Mar. 1953, *Drummond & Hemsley* 1505 !
ZANZIBAR. Zanzibar Island, Bukubu, Oct. 1950, *R. O. Williams* 73 ! ; Zingwe-Zingwe, Jan. 1929, *Greenway* 1095 ! ; Pemba, near Ngezi Forest, Aug. 1929, *Vaughan* 612 ! & Shengejuu–Pandani, Feb. 1929, *Greenway* 1500 !
DISTR. U2, 4 ; K?3, 4, 5 ; T1–8 ; Z ; P ; throughout tropical Africa ; also in tropical Asia, Australia, and America ; naturalized in Madagascar
HAB. Damp places, often growing in water ; sea level to 2400 m.

SYN. *P. serrulatum* Lag., Gen. & Sp. Pl. Nov. : 14 (1816) ; F.T.A. 6 (1) : 107 (1909).
          Type : from Spain
       P. erythropus Dammer in Engl., P.O.A. C : 170 (1895). Type : Tanganyika, Tabora District, Ugunda, Igonda [not " Uganda, Evonda "], *Böhm* 34 (B, holo. !)
       P. serrulatum Lag. var. *angustifolium* Peter in Abh. Ges. Wiss. Göttingen, n.f. 13 (2) : 52 (1928). Type : Tanganyika, Kigoma District, Ujiji, *Peter* 37026 (B, holo. †)
     ?*P. serrulatum* Lag. var. *donii* sensu Peter, in Abh. Ges. Wiss. Göttingen, n.f., 13 (2) : 52 (1928) *non* Hook. f.
       P. serrulatum Lag. var. *bequaertii* De Wild., Pl. Bequaert. 5 : 265 (1931). Type· Belgian Congo, Kivu District, Rutshuru, *Bequaert* 6238 (BR, holo.)
       P. leuconeuron Peter, F.D.O.-A. 2 : 189 (1932) & anh. : 20 (1938), ex descr. & tab.
          Type : Tanganyika, Pare Mts., Mbaga, *Peter* 8961b (B, holo. †)
       P. serrulatum Lag. forma *polystachyum* Peter, F.D.O.-A. 2 : 189 (1932) &

anh.: 20 (1938), ex descr. Type: Tanganyika, Uzaramo District, between Dar es Salaam and Lake Buharati, *Peter* 44621 (B, holo. †)

VARIATION. The nuts are usually trigonous, but are lenticular in *P. serrulatum* var. *bequaertii.* Between the two extremes are intermediates, best described as " obscurely trigonous." *P. erythropus* is apparently a form perhaps induced by habitat, with shorter more dense racemes with bracts contiguous throughout, and with perianths 3 mm. long : it perhaps shows an extreme in raceme density for the species, but forms approaching in this character occur.

8. **P. senegalense** *Meisn.*, Monogr. : 54 (1826) & in DC., Prodr. 14 : 123 (1856) ; F.T.A. 6 (1): 111 (1909) ; F.P.N.A. 1 : 119 (1948) ; F.C.B. 1 : 416 (1948) ; F.W.T.A., ed. 2, 1 : 141 (1954). Type : Senegal, *Perrottet* (G–DC, holo.)

An erect, robust perennial up to 3 m. tall. Stem often rooting at the basal nodes, varying from glabrous to thickly and softly white-tomentose, often obscured by the ocreae. Ocreae up to 3·5 cm. long, membranous, reddish-brown, truncate, glabrous or covered in variable thickness with a cobwebby tomentum. Leaves large, petiolate, oblong-lanceolate, up to 27·5 × 6·5 cm. ; with a long, acute or acuminate apex, basally narrowed to the petiole ; glabrous throughout (or with short spiny bristles on the sub-surface midrib) to densely white-tomentose on both faces or predominantly below, the undersurface covered with small, yellowish glands. Petioles up to 2·5 cm. long, glabrous to tomentose. Inflorescence a branched, leafless panicle of one to several, densely flowered spiciform racemes, the peduncles arising in pairs (more rarely in threes) from the short, cup-like, upper ocreae. Peduncles covered with orange glands, puberulent varying to white-tomentose. Bracts broadly ovate, truncate, glabrous or pubescent, usually with at least a few orange glands. Perianth 3 mm. long, rose-pink, greenish or white, dotted with glands. Tepals 4, longer than the tube, elliptic-oblong, the outer pair a little broader than the inner. Nut lenticular with dimpled faces, 2·5 mm. long (apical beak 0·5 mm.), shining. This species exudes a yellow substance from its glands on to drying papers when under pressure. Fig. 4/5 and 6, p. 23.

forma **senegalense**. Type as for species

Whole plant glabrous or essentially so. Leaves green, lush.

UGANDA. Lango District : Ayer, May 1941, *A. S. Thomas* 3881 ! ; Ankole District : Mitoma, March 1935, *Purseglove* 598 ! ; Mengo District : Entebbe, Sept. 1922, *Maitland* 209 !
KENYA. Naivasha, Aug. 1934, *Turner* 6706 ! ; Embu District : Emberre, Itabwa, Aug. 1922, *M. D. Graham* 2102 !; Kisumu–Londiani District: Muhoroni, *Battiscombe* 66 !
TANGANYIKA. Bukoba District : Bugufi, Jan. 1936, *Chambers* K22 ! ; Ufipa District : Lake Kwela, Mar. 1950, *Bullock* 2645 (growing with the holotype of f. *albotomentosum*); Rufiji District : Mafia Is., Kerongwe, Aug. 1937, *Greenway* 5145 !
DISTR. **U**1–2, 4 ; **K**3–5 ; **T**1–7 ; widely spread through tropical Africa ; also in Madagascar, South Africa and Egypt
HAB. By lakes & riversides and in other damp places, often growing in water ; wide-spread and locally common ; 400–3000 m.

SYN. *P. senegalense* Meisn. var. *usambarense* Dammer in Engl., P.O.A. C : 170 (1895). Type : Tanganyika, ? Lushoto District, by R. Umba, *Holst* 403 (B, holo.)
    P. sambesicum Schuster, in Bull. Herb. Boiss., sér. 2, 8 : 708 (1908). Type : Portuguese East Africa, R. Zambezi, *Menyhart* 710 (Z, holo. !)
    P. tanganyikae Schuster in Bull. Herb. Boiss., sér. 2, 8 : 709. Type : Tanganyika, Mpanda District, Karema, *Böhm* 74a (Z, lecto. !)
    ?*P. tanganyikae* Schuster var. *ciliatum* Peter in Abh. Ges. Wiss. Göttingen, n.f., 13 (2): 53 (1928). Type : Tanganyika, Masai District, Engaruka, *Peter* 42816 (B, holo. †)
    [?*P. glabum* sensu Peter, l.c. 53 (1928), *non* Willd. No specimens of Peter's seen.]
    ?*P. sambesicum* Schuster var. *angustifolium* Peter, F.D.O.-A. 2 : 195 (1932) &

anh.: 21 (1938). Type: Tanganyika, Buha District, between Kasulu and Kivumba, *Peter* 37547 (B, holo. †)

forma **albotomentosum** *R. Grah.* in K.B. 1956: 258 (1956). Type: Tanganyika, Ufipa District, Lake Kwela, *Bullock* 2645 (K, holo.!)

Stems, leaves, ocreae and peduncles thickly covered with white or whitish tomentum. Leaves white or whitish on both sides or more ashen above.

UGANDA. Karamoja District : Mt. Debasien, Napyenenya, Jan. 1936, *Eggeling* 2575 ! ; Ankole District : Mitoma, Mar. 1936, *Purseglove* 597 ! ; Busoga District : Bugaya, Nov. 1949, *Jameson* 76 !
KENYA. Nairobi, Sept. 1915, *Dowson* 297 ! ; S. Kavirondo District : Sasi River, June 1945, *Glasgow* 45/10
TANGANYIKA. Moshi District : Kibosho, Dec. 1930, *Sanders* 26 ! ; Rungwe District : Mwankinja, Feb. 1954, *Semsei* 1605 ! and Tukuyu, Sept. 1932, *Geilinger* !
ZANZIBAR. Zanzibar Island, unlocalized, *Bojer* !
DISTR. **U**1–4 ; **K**4, 5 ; **T**2, 6–8 ; **Z** ; the general distribution is that of forma *senegalense,* but records from Kenya are few
HAB. As for forma *senegalense;* available records give an altitude range of 45–1650 m., which is lower than that of forma *senegalense,* but it is not known whether this is due to the chances of collecting, or is perhaps indicative of a preference for lower altitudes

SYN. *P. lanigerum* R. Br. var. *africanum* Meisn. in DC., Prodr. 14 : 117 (1856) ; F.P.N.A. 1 : 122 (1948) ; F.C.B. 1 : 420 (1948) ; F.W.T.A., ed. 2, 1 : 141 (1954). Type : South Africa, *Burchell* 2281 (? G–DC, syn.)
[*P. lanigerum* sensu F.T.A. 6 (1) : 109 (1909) & Fl. Cap. 5 (1) : 468 (1912); *non* R. Br.]
P. bussei Peter in Abh. Ges. Wiss. Göttingen, n.f. 13 (2) : 53 (1928). Type : Tanganyika, Kilosa District, near Kitete, *Peter* 32768 (B, holo. †)
P. lanigerum R. Br. var. *salicifolium* Peter in F.D.O.-A. 2 : 190 (1932.) & anh. : 21 (1938). Type: Tanganyika, Buha District, Lake Manyoni near Mbirira, *Peter* 37803 (B, holo. †)
?*P. senegalense* var. *subciliatum* Peter in F.D.O.-A. 2 : 190 (1932) & anh.: 21 (1938). Type : Tanganyika, Tanga District, Udigo, *Peter* 23853b (B, holo. †)

VARIATION. Intermediates between the two forms commonly occur, showing different degrees of hairiness : *Drummond & Hemsley* 2082 ! has whitish tomentum on young leaves only ; *Bullock* 2342 ! has leaves with thinly hairy, green upper surfaces, and white tomentose subsurfaces becoming glabrous. Another probable intermediate, according to the description, is *P. lanigerum* forma *decalvans* Peter (Type : Tanganyika, Mkata, *Peter* 32366 (B, holo. †))

NOTE. On the shape of the fruit, as depicted in its figure, *P. bussei* must be related to *P. senegalense,* and, from the description, to forma *albotomentosum.* But in the absence of any authentic specimens the exact status of Peter's species must remain in doubt : in particular, multisetose ocreae are unusual in *P. senegalense* (agg.).

9. **P. amphibium** *L.*, Sp. Pl.: 361 (1753) ; Meisn., Prodr. Polygon. : 67 (1826), and in DC., Prodr. 14 : 115 (1856) ; Syme in Smith, Engl. Bot., ed. 3, 8 : 77, tt. 1241–2 (1868). Type : from Europe

A variable perennial, with two distinct growth forms depending upon habitat.

Floating in water. Stems weak, glabrous, green to greenish-brown, trailing, rooting at the nodes in the water. Ocreae 0·8–2 cm. long, scarcely tearing. Leaves floating, on weak petioles 2·5–8 cm. long ; oblong or oblong-lanceolate, up to 17·5 × 4·5 cm. ; apically rounded, basally rounded, truncate or cordate ; glandular or not. Inflorescence a rather short, dense raceme of reddish-pink flowers, 3·5–4·5 × 1–1·5 cm., simple or with another flowering peduncle arising from the topmost node. Pedicels 2 mm. long or less. Perianth 4–5 mm. long. Tepals 5, ovate, ± 3 mm. long. Stamens 5, included or exserted. Styles 2, united for $\frac{1}{2}$–$\frac{2}{3}$ their length. Nut biconvex-lenticular, dark brown, shining, 2·25×2 mm.

Terrestrial or growing on drying mud. Stems erect, rooting only at the lower nodes. Leaves glabrous or pubescent, glandular or not, narrowly lanceolate, commonly 16·5 × 3 cm., apically acute, basally rounded, truncate or cordate, subsessile or on petioles up to 1 cm. long. Inflorescence characters as above.

There is inter-variability between the two growth-forms : the single specimen from our area is the aquatic form, and is eglandular.

KENYA. Aberdare Mts., Kinangop, Apr. 1938, *Chandler* 2339 !
DISTR. **K**3–4 ; North America, Europe, temperate Asia, and North Africa ; probably introduced into Kenya and South Africa
HAB. Ponds or waterways ; also in wholly terrestrial habitats, and as a weed of disturbed ground

10. **P. limbatum** *Meisn.* in DC., Prodr. 14 : 123 (1856) ; F.T.A. 6 (1) : 108 (1909) ; F.C.B. 1 : 414 (1948) ; F.W.T.A., ed. 2, 1 : 140, fig. 51 (1954). Types : ? Senegal River, *Lelièvre* (B, syn. †) & Nile Delta, *Ehrenberg* (B, syn. †)

An erect perennial herb. Stems branched, green or red, basally decumbent and rooting at the nodes, usually with ascending, appressed hairs in the upper parts, or more rarely ± glabrous throughout or thickly covered with long white hairs. Ocreae glabrous to hispid, membranous, brown, terminating in a spreading, green, foliaceous, undulating limb, 1–9 mm. broad, with a ciliate margin. Leaves sessile or subsessile, variable in shape but often narrow-lanceolate, sometimes dotted with glands, 8–11 (–15) × 0·8–1·8 (–3·5) cm., apically acute, narrowed to the base, sometimes crisped on the margins, usually pubescent, sometimes covered with silky hairs on both faces. Inflorescence a dense raceme, 2–6 cm. long, on long, pubescent peduncles arising in pairs from the uppermost ocreae (singly *fide* Meisn.). Bracts, at least the lower ones, usually pubescent with ciliate margins and a terminal fringe of cilia, but varying from glabrous and terminally non-ciliate to densely hirsute with long, silky hairs. Perianth pink 3·5–4·5 mm. Tepals 5, ovate-oblong, 2·5–3 mm. long. Styles 2, connate for approximately half their length. Nut ± rounded in outline, 2·5 × 2·5 mm., dark red-brown to black, shining, biconvex-lenticular. Fig. 4/7, p. 23.

UGANDA. Teso District : Soroti, Sept. 1954, *Lind* 330 !
TANGANYIKA. Moshi District : Boloti swamp, Aug. 1928, *Haarer* 1473 ! ; Iringa District : Little Ruaha River, Mar. 1932, *Lynes* P.r.63 ! ; Njombe District : R. Ruhudje, near Lupembe, May 1931, *Schlieben* 827 !
DISTR. **U**3 ; **T**2, 7–8 ; from Egypt to Rhodesia ; and west from northern Nigeria to Senegal ; also in tropical Asia
HAB. Damp places, often growing in water ; 1000–1500 m.

VARIATION. From the scanty East African material available for consultation it appears that this species is very variable in the degree and nature of hairiness. In *Haarer* 1473, the leaves and bracts are covered with silky, white hairs, those on the stem and ocreae being very thick, 3–4 mm. long and spreading at 45°. In *Lind* 330, the ocreae have a thick covering of brown hairs, 2–3 mm. long and spreading at all angles, while the leaves are closely covered throughout with very short setae, and the bracts with stiff bristles. The more usual form appears to be that with thinly pubescent leaves with stiffer hairs on the midribs, but until further material is available it is difficult to dogmatize. It is possible that *Haarer* 1471 and *Lind* 330 represent different taxa.

NOTE. *P. schinzii* C. H. Wright (in K.B. 1909 : 187 (1909). Type : South West Africa, Amboland, *Schinz* 499 (K, holo. !)) *non* Schuster is a thinly hairy plant with crisped leaf margins apparently closely allied to *P. limbatum*, and perhaps conspecific with it, but for the reason set out in the foregoing paragraph it is considered that the matter is too uncertain at the moment to justify synonymy.

11. **P. pulchrum** *Blume*, Bijdr. : 530 (1826) ; F.C.B. 1 : 419, fig. 40 (1948) ; F.P.N.A. 1 : 122 (1948) ; F.W.T.A., ed. 2, 1 : 141 (1954). Type : Java, Batavia, collector and location uncertain

A stout softly hirsute perennial, 1 m. tall or more. Stems branched, basally creeping and rooting at the nodes, brownish, covered with appressed, ascending, brown hairs, varying to subglabrous. Ocreae light brown, membranous, the lower ones up to 3·5 cm. long, setose becoming glabrous with age, truncate with a terminal fringe of hard bristles which may equal the length of the tube. Leaves shortly petioled, variable in size and shape, commonly broadly ovate-acuminate but varying to narrow-lanceolate, elliptic-lanceolate, oblong-lanceolate, or oblong-elliptic, broadest usually below the middle, thence narrowed to a long, drawn-out, acute or acuminate apex, and more shortly narrowed to the base, the mature ones commonly 12–18 × 2·5–5·5 cm., pubescent above, covered below (sometimes only thinly) with a soft, grey-white tomentum, midrib often prominent below, with many arcuately ascending, ± prominent lateral veins commonly 2–3 mm. apart. Petioles not or seldom exceeding 1·2 cm., the upper leaves often subsessile. Inflorescence a dense spiciform raceme, 5–8 cm. long and up to 1 (–2) cm. broad, usually leafless. Peduncles stout, often in pairs from the uppermost ocreae and rising somewhat geniculately, tomentose with ascending, whitish hairs. Bracts reddish, rather thinly setose or glabrous, ovate, truncate, terminally fringed. Pedicels exceeding the bracts by 1–2 mm. Perianth pink (3–) 4–5 mm. long. Tepals broadly oblong-elliptic, 2–3 mm. long. Styles 2, united at or shortly below the middle, variable in length. sometimes exserted. Nut rather unequally biconvex-lenticular, shining, black, 2–3 mm. long, broadly ovoid-ellipsoid to spherical, rarely ± trigonous. Fig. 4/3, 4 and 8, p. 23.

UGANDA. Ankole District : Igara, Feb. 1936, *Purseglove* 590! ; West Nile District : Maracha, Dec. 1939, *Hazel* 401! ; Teso District : Agu Swamp, Sept. 1932, *Chandler* 976!
KENYA. Nairobi, June 1933, *Napier* 2659! ; Embu District : Emberre, Itabwa, Mar. 1932, *Sunman* 2220! ; S. Nyeri District : Thiba River, Nov. 1939, *Copley in Bally* 390!
TANGANYIKA. Mwanza, shore of Lake Victoria, Aug. 1932, *Rounce* 1981! ; Pangani District : south bank of Pangani River, between Hale and Makinyumbe, July 1953, *Drummond & Hemsley* 3130! ; Mbulu District : Babati, Oct. 1925, *Haarer* 18b!
ZANZIBAR. Zanzibar Is., Kinysaini, Jan. 1929, *Greenway* 1118! and without locality, *Vaughan* 1428!, 1102!, 1878!, & *Kirk*!
DISTR. U1–4 ; K3, 4 ; T1–8 ; Z ; tropical and subtropical Africa and Asia
HAB. Damp places, river-sides, papyrus-swamps, etc., sometimes in brackish water ; sea level–1950 m.

SYN. [*P. ochreatum* sensu Houtt., Handl. 8 ; 467, t. 49/1 (1777), *non* L.]
P. tomentosum Willd., Sp. Pl. 2 (1) : 447 (1800) ; Meisn. in DC., Prodr. 14 : 124 (1856) ; F.T.A. 6 (1) : 110 (1909) ; F.W.T.A. 1 : 120 (1927), *non* Schrank (1789), *nom. illegit.* Type : Ceylon, collector and location uncertain
P. tomentosum Willd. var. *glandulosum* Meisn. in Linnaea 14 : 484 (1840). Type : South Africa, unlocalized, *Drège*
P. tomentosum Willd. var. *limnogenes* (Vatke) Hiern, Cat. Welw. Afr. Pl. 4 : 905 (1900). Type : Angola, Huilla, *Welwitsch* 5362 (K, lecto. !)
?*P. tomentosum* Willd. var. *blepharanthum* Peter in Abh. Ges. Wiss. Göttingen, n.f. 13 (2) : 53 (1928). Type : Ruanda-Urundi, between Malagarasi River and Mgoni, *Peter* 46312 (B, holo. †)
?*P. tomentosum* Willd. var. *eciliatum* Peter, l.c. Type: Tanganyika, Kigoma District, Uvinza, *Peter* 36229 (B, holo. †)
P. tomentosum Willd. forma *angustifolium* De Wild., Contrib. Fl. Katanga, Suppl. 3 : 103 (1930). Type : Belgian Congo, Katanga, road to Étoile du Congo Mine, *Quarré* 390 (BR, holo. !)
?*P. tomentosum* Willd. var. *brevipilum* Peter in F.D.O.-A. 2 : 194 (1932) & anh.: 21 (1938). Type : Tanganyika, Pangani District, islands in Pangani River near Hale, *Peter* 4502c (B, holo. †)
?*P. tomentosum* Willd. var. *angustifolium* Peter, F.D.O.-A. 2: 194 (1932) & anh.: 22 (1938). Type : Tanganyika, Pangani District, Pangani River near Hale, *Peter* 40533 (B, holo. †)

NOTE. *P. pulchrum* is offered here as being almost identical with Asian material, the differences between the African and Asian races being only slight, and there being considerable dove-tailing owing to the variability of the species. Broadly, Asian examples have leaves tending to be widest nearer the middle and with rather more rounded and slightly larger nuts. The American *P. acuminatum* H.B.K. can perhaps be regarded as a closely allied geographical race of the same species, though the difference in the orientation of the cotyledons may indicate a more distant relationship ; but from herbarium specimens seen it would appear that *P. pulchrum* has broader leaves, usually tomentose beneath with soft hairs on the midrib (in *P. acuminatum* they are glabrous or pubescent with bristly hairs on the midrib) and has rather less congested racemes.

This species is very variable, especially in leaf shape, but the softly hairy peduncles and leaf subsurfaces, the larger, often more clearly exserted flowers, and the stoutness of the plant usually serve to distinguish it from the next species without difficulty. Narrow, linear leaved forms are apparently commoner in territories west and south of our area.

12. **P. setosulum** *A. Rich.*, Tent. Fl. Abyss. 2 : 227 (1851). Type : Ethiopia, Shire, *Dillon & Petit* (P, holo. !)

A variable, ± robust perennial. Stems basally decumbent, rooting at the nodes, often reddish, glabrous or nearly so, up to 1 m. or more tall. Ocreae 1–3 cm. long, membranous, thinly to (more commonly) thickly covered with closely ascending bristly hairs 1–2·5 mm. long, terminally truncate and (at least when young) with a terminal fringe of bristles (0·3–) 0·5–1·1 cm. long. Leaves shortly petiolate, variable in size and shape, commonly 11–12 × 2–2·5 cm., ovate-acuminate, ovate-lanceolate, or elliptic-lanceolate, broadest usually below the middle, sometimes (as in holotype) with a black blotch in the form of a reversed V in the centre, apically acute, basally narrowed to the petiole, usually setose on the margin and midrib of the upper surface, more thickly so on the midrib and lateral nerves below and thinly so on the lamina, but varying to almost glabrous ; lateral nerves of the undersurface often alternately prominent and obscure, the prominent ones 0·4–1·0 cm. apart. Petioles up to 1 cm. long, commonly less. Inflorescence simple or branched, racemose, moderately stout, the peduncles 2–3 together and arising often at an acute angle, 2·4–6·0 cm. long, ± densely flowered. Peduncles glabrous, more rarely setose. Bracts usually with a terminal fringe of short bristles 1–2 (–5) mm. long, otherwise glabrous, sometimes dotted with a few orange glands. Pedicels short, rarely exceeding the bracts by more than 1·25 mm. Perianth white to deep pink, sparsely or not glandular, 2·75–3·0 mm. long. Tepals 5, ovate-oblong, 2–2·5 mm. long, rarely more. Styles 2, 1·75–2 mm. long, united for 0·5–0·75 mm. Nut 2–3 (–3·25) mm. long, smooth, shining, evenly biconvex-lenticular or with one side swollen to become obscurely trigonous, rarely ± acutely trigonous. Fig. 4/1 and 2, p. 23.

UGANDA. Karamoja District : Mt. Debasien, Namojongotyang, *Eggeling* 2635 ! ; Kigezi District : Kabale-Mbarara road, Mar. 1952, *Norman* 93 !* ; Toro District : Ruwenzori, Mubuku Valley, Aug. 1933, *Eggeling* 1258 !

KENYA. Nairobi, Sept. 1949, *Bogdan* 1136 ! ; Kericho District : Itare river, Kipsoi Mill, Oct. 1940, *Copley in Bally* 1191 ! ; Naivasha District : Kedong, Mt. Margaret Estate, June 1940, *Bally* 908 !

TANGANYIKA. Masai District : Ngorongoro crater-floor, Apr. 1941, *Bally* 2312 ! ; Moshi District : Lyamungu, Feb. 1942, *Wallace* 1050 ! ; Lushoto District : near Lushoto, Mkuzi, Apr. 1953, *Drummond & Hemsley* 2155 !*

DISTR. U1–4 ; K3–5 ; T2–4, 6–8 ; range uncertain outside tropical and subtropical Africa

HAB. Damp places, sometimes growing in water ; especially in upland areas from 1050–2670 m.

SYN. *P. poiretii* Meisn. var. *latifolium* Dammer in P.O.A. C. : 170 (1895). Type : Tanganyika, W. Usambara Mts., Kwa Mshusa, *Holst* 9028 (K, holo. !)
       P. barbatum L. var. *fischeri* Dammer in P.O.A. C : 169 (1895). Type : Tanganyika, unlocalized, *Fischer* 231 (B, holo. !)

* With blotched leaves.

FIG. 4. *POLYGONUM SETOSULUM*—**1,** flowering branch, × 2/3 ; **2,** spotted leaf, × 2/3 ; *P. PULCHRUM*—**3,** flowering branch, × 2/3 ; **4,** fruit showing front and convex sides, × 3 ; *P. SENEGALENSE* forma *SENEGALENSE*—**5,** ocrea, × 2/3 ; **6,** fruit showing front and dimpled sides, × 3 ; *P. LIMBATUM*—**7,** ocrea, × 2/3 ; *P. PULCHRUM*—**8,** flower opened to show petaloid tepals, stamens, immature fruit, and 2 styles connate below, × 3.

P. nyikense Baker in K.B. 1897 : 280 (1897) ; F.W.T.A., ed. 2, 1 : 142 (1954). Type : Nyasaland, Nyika Plateau, *Whyte* (K, holo. !)

[*P. barbatum* sensu F.T.A. 6 (1) : 109 (1909) ; Fl. Cap. 5, 1 : 467 (1912) ; Fl. Madag. 65 : 10 (1953) ; *non* L.]

[*P. acuminatum* sensu F.T.A. 6 (1) : 112 ; Fl. Madag. 65 : 12 (1953) ; F.C.B. 1 : 418 (1948) ? pro parte; *non* H.B.K.]

P. mildbraedii Dammer in Z.A.E. : 202 (1911) ; F.P.N.A. 1 : 120 (1948) ; F.C.B. 1 : 418 (1948). Type : Ruanda-Urundi, Rugege, *Mildbraed* 931 ? B, holo.)

?*P. hololeion* Peter in Abh. Ges. Wiss. Göttingen, n.f. 13 (2) : 53 (1928). Type : Tanganyika, Mpwapwa District, Gulwe, by Lake Kimagai, *Peter* 32874 (B, holo.†)

?*P. holotrichum* Peter, l.c. 52. Type : Tanganyika, E. Usambara Mts. between Derema and Msituni, *Peter* 21518 (B, holo.†)

?*P. hydrophilum* Peter, l.c. 52. Type : Tanganyika, Kigoma District, Uvinza, *Peter* 36447 (B, holo.†)

P. quarrei De Wild., Contrib. Fl. Katanga, Suppl. 3 : 105 (1930). Type : Belgian · Congo, Kufubu, *Quarré* 268 (BR, syn. !)

NOTE. *P. barbatum* L., with which this species has been confounded, differs in its denser racemes of smaller flowers, and in its nuts which apparently do not exceed 2 mm. in length and appear always to be rather sharply trigonous. The two species may, however, have a geographical consanguinity. *P. acuminatum* H.B.K. differs in its longer leaves, stouter, more congested inflorescences, and in that its nut is apparently always lenticular.

The frequency of leaf blotches in *P. setosulum* cannot be accurately assessed, as in pressed examples the mark usually disappears unless drying is carried out with particular care.

No specimens of *P. hololeion* Peter have been seen. The figure shows a plant with narrow leaves like those of *P. salicifolium*, but the flowers, 3–3·5 mm. long (*in descr.*) are too large for the latter, and it is probable that Peter's species is a narrow-leaved form of *P. setosulum*. The same applies perhaps to his *P. hydrophilum*.

P. barbatum var. *fischeri* is a narrow-leaved form, with long racemes (up to 10 cm.) and clearly trigonous fruit.

13. **P. persicaria** *L.*, Sp. Pl. : 361 (1753) ; Meisn., Prodr. Polygon. : 68 (1826) & in DC., Prodr. 14 : 117 (1856) ; Syme in Smith, Engl. Bot., ed. 3, 8: 74, t. 1237 (1868). Type : from Europe

An annual, largely glabrous herb, up to ± 1 m. tall but usually less. Stems branched, green or red, glabrous or subglabrous, often basally decumbent and rooting at the nodes. Ocreae 1–1·7 cm. long, appressed when young, terminally fringed with usually short bristles. Leaves petiolate or sometimes subsessile, often with a black blotch centrally on the upper surface, lanceolate or elliptic-lanceolate, broadest usually below the middle, up to 10 × 2·2 cm., gradually narrowed to an acute apex and to the petiole, usually glabrous, sometimes with appressed pubescence on the upper surface and with scattered hairs on the midrib below. Petioles up to 1 cm. long. Inflorescence a branching, open, usually leafy panicle, with terminal, rather short racemes 1–3 cm. long. Bracts pink, green or white, sometimes hyaline, fringed with setae, glabrous. Flowers pink, green or whitish green, dense. Peduncles slender, usually eglandular. Perianth c. 3 mm. long. Tepals 5, eglandular or occasionally with a few glands, broadly ovate-elliptic or nearly rounded, ± 2 mm. long. Styles 2 or 3, united for half their length. Nut dark reddish-brown, bluntly trigonous, 2 (–3) mm. long, very shortly beaked.

UGANDA. Busoga District : Mutai Forest Reserve, about 17 km. N. of Jinja, Nov. 1952, *Wood* 565 !

DISTR. U3 ; probably introduced ; temperate regions in the northern hemisphere ; very common in western Europe

HAB. Damp places, and as a weed of disturbed ground; 1100 m.

14. **P. baldschuanicum** *Regel* in Act. Hort. Petrop. 8 : 684 (1883). Type : Turkestan, Baldschuan, *Regel* (K, iso. !)

A decorative woody, climbing and twining herb, glabrous throughout. Stems brownish red. Leaves in clusters at the nodes of the main stems, ovate, apically acute to rounded, basally hastate to cordate and very shortly cuneate to the petiole, up to 7·5 × 4·5 cm., sometimes as broad as long. Petioles usually long, 2–6 cm. Inflorescence a much-branched, diffuse, terminal panicle of many slender racemes 3–10 cm. long. Bracts readily tearing. Flowers whitish 3–4 in each cluster. Pedicels filiform, 3–12 mm. long. Tepals 5, the inner two obovate, the outer three ovate, winged, the wings decurrent down the pedicel. Stamens 8. Styles 3, ± sessile. Nut black, smooth, shining, trigonous, ± 5 mm. long.

KENYA. Meru, June 1951, *Hancock* 60 !
DISTR. **K4** ; a native of Turkestan ; widely planted for decoration in gardens
HAB. " In hedges and fields. Climber in gardens." No doubt an escape, which may, however, spread locally.

15. **P. convolvulus** *L.*, Sp. Pl. : 364 (1753) ; Meisn. in DC., Prodr. 14 : 135 (1856) ; Syme in Smith, Eng. Bot., ed. 3, 8 : 61 (1868). Type : from Europe

A procumbent, or twining and climbing annual herb, largely glabrous. Stems weak, green or greenish brown, almost glabrous, rather sharply edged and sometimes mealy on the edges. Ocreae brown, membranous, glabrous or shortly pubescent, sometimes with scattered glands at the base, readily tearing, ± 3 mm. long. Leaves petiolate, ovate-acuminate, basally sagittate with a broad basal sinus, 4–6 × 2·5–4 cm., glabrous, rather mealy below. Petioles slender, usually mealy, up to 4 cm. long. Inflorescence a slender, interrupted raceme, often leafy almost to the apex. Pedicels ± 1·75 mm. long. Perianth ± 4 mm. long. Tepals 5, green with white edges, 1·5–2 (–3) mm. long, mealy on the outside, oblong, the inner two flat, the outer three keeled or slightly winged (but if so the wings not decurrent down the pedicels). Nut black, trigonous, matt or shiny on the edges, 3–4 × 2·25–2·5 mm.

KENYA. Trans-Nzoia District : about 19 km. W. of Kitale, Sept. 1948, *Bogdan* 1950 ! ; Nakuru District : Ol Joro Orok, Dec. 1955, *Peers* C11 ! ; Kiambu District : Kabete, Dec. 1948, *Nattrass in Bally* 6515 !
DISTR. **K3, 4** ; temperate Europe and Asia, N. Africa (? introduced) ; elsewhere introduced
HAB. A weed in cultivated ground, also in hedgerows, etc. ; introduced in our area as a weed in maize and other crops, and becoming locally frequent in the Nakuru and Eldoret districts of Kenya; 1800–2200 m.

Uncertain species

P. tumidum *Del.*, Fl. Aegypt. Illustr. : 60 (1813). This species is referred to by Dammer (P.O.A. C : 170 (1895)) and by Peter (in F.D.O.-A. 2 : 196 (1932)) as occurring at Karema, in Mpanda District, on Lake Tanganyika. No specimens are cited, and any retained were probably destroyed. Apart from this, the exactitudes of Delile's species have not been established, and the name may be referable to *nomina nuda*. The identity of the plant recorded therefore remains in doubt.

5. **FAGOPYRUM**
Mill., Gard. Dict., abridg. ed. 4, 1 (1754)

F. esculentum *Moench.*, Meth.: 290 (1794).

An erect, glabrous or puberulous annual herb, with sagittate-cordate leaves, pink or white flowers arranged in axillary and terminal umbels or racemes, and with a very accrescent, brown, matt, sharply trigonous nut (5–6 mm. long) with narrowly winged edges.

UGANDA. Kigezi District : Nyebeya, Oct. 1940, *Eggeling* 4138 !
KENYA. Nairobi, weed in the Coryndon Museum garden, June 1947, *Turner in Bally* 5035 !
TANGANYIKA. W. Usambara Mts., Lushoto, *Peter* 3931 (B, †)
DISTR. U2 ; K4 ; T3 ; native in Central Asia, grown as a cover-crop at Kampala, elsewhere an introduced weed

SYN. *Polygonum fagopyrum* L., Sp. Pl. : 364 (1753). Type : originally from Asia, sheet 510/37 in *Herb. Linnaeus* (LINN, lecto.. !)

6. OXYGONUM

Burch., Trav. 1 : 548 (1822) ; R. Grah. in K.B. 1957 : 145 (1957)

Polygamous, heterostylous, annual or perennial herbs or more rarely shrubs. Ocreae with or without a terminal fringe of setae. Flowers borne in axillary, leafless, often triquetrous, rather slender and elongated spiciform racemes. Bracteoles 2, conjoined. Male flowers with a very short tube and 4–5 petaloid tepals. Hermaphrodite flowers tubular, the tepals marcescent, the tube very accrescent around the ovary. Stamens 8, in 2 series ; 5 outer adnate to the tepals near their bases, 3 inner with flattened bases forming a contiguous ring around the base of the style. Styles 3, free or conjoined below ; stigmas capitate. Fruiting perianth fusiform, conical, ampulliform or lanceolate-ampulliform, trigonous or not, unarmed, or with 3 (–9) spreading prickles of varying size and development at the centre or below, or at the base, rarely with 3 additional prickles or teeth on the angles of the lower half, sometimes winged.

A genus confined to Africa, except for two species, one of which is an endemic in Madagascar.

Leaves large, obovate-oblong, broadly ovate, or broadly oblong-elliptic, 7·5–10·5 × 4·5–5·5 cm., basally attenuate, often long-attenuate to the petiole ; the margin entire or slightly uneven, not lobed .   .   .   .   1. *O. buchananii*
Leaves (except in the deeply-lobed *O. lobatum*) not as large as the preceding and usually much smaller, deltoid, ovate, lanceolate, elliptic, or linear, but very variable, the margin entire or often at least sinuous and sometimes deeply lobed, basally rounded, hastate, sagittate, or attenuate :
  Fruit unarmed (or with 3 teeth or rudimentary teeth at base) :
    Fruit ribbed and wrinkled, unarmed, lanceolate-ampulliform ; leaves linear or narrowly elliptic-lanceolate, entire or faintly sinuous.   2. *O. dregeanum* var. *strictum*

    Fruit usually with 3 teeth or rudimentary teeth at or near the base, not exceeding 5 (–6) mm. in length, pubescent, conical ; leaves linear to lanceolate, or entire or lobed (often unevenly). (If fruits ± 10 mm. long, see No. 13)   3. *O. delagoense*
  Fruit with prickles in the centre or in the lower half, fusiform or conical, occasionally winged :
    Leaves basally sagittate, closely white-tomentose ; pedicels scarcely exceeding the bracts. (If leaves with sagittate bases but with long, filiform pedicels, then see No. 14)   .   .   4. *O. sagittatum*

Leaves not basally sagittate (but see No. 14) :

Leaves linear or narrowly linear-lanceolate, the margin entire or slightly wavy but not lobed :

Fruiting pedicels ± erect, exceeding the bracts by 2–5 mm. ; creeping perennial herb ; stems rather woody . .   5. *O. salicifolium*

Fruiting pedicels spreading, filiform, exceeding the bracts by (6–) 8–10 mm. ; leaves clustered ; suberect herb . . . 6. *O. subfastigiatum*

Leaves variable, but neither linear, nor narrowly linear-lanceolate :

Fruiting pedicels noticeably long, exceeding the bracts by more than 1 cm. :

Leaves lanceolate to elliptic-lanceolate, basally long narrowed and decurrent to the petiole . . . . 7. *O. schliebenii*

Leaves ovate or ovate-deltoid, basally rounded or abruptly narrowed to the petiole . . . . . . 8. *O. leptopus*

Fruiting pedicels short, exceeding the bracts by not more than 8 mm. :

Fruiting pedicels exceeding the bracts by 5–8 mm. ; fruit 8 mm. long ; leaves deeply lobed, shortly puberulous . 9. *O. hirtum*

Fruiting pedicels exceeding the bracts usually by less than 5 mm. :

Leaves elliptic, entire, evenly narrowed to each end, not exceeding 5·5 × 2 cm. :

Petioles ± 1·5 cm. long ; raceme-stalk 0·5–1 mm. thick ; fruit 7–8 mm. long, with 6 (–9) prickles . . 10. *O. ellipticum*

Petioles scarcely or not exceeding 6 mm. in length ; fruit with 3 prickles only :

Fruit 9–10 mm. long ; leaves hairy to glabrous ; raceme-stalk 0·75–2 mm. thick ; short styles 2 (–1·5) mm. long . 11. *O. magdalenae*

Fruit 5 mm. long ; leaves glabrous or ± so ; short styles 1 mm. long . . . . . 16. *O. stuhlmannii*

Leaves hastate, deltoid, ovate, or lanceolate, ± entire to deeply incised with rounded lobes ; if elliptic or ± so, then the margin uneven :

Stigmas of long-styled flowers included, not or only a little protruding beyond the tips of the anthers ; those of short-styled flowers reaching to or nearly to the bases of the anthers ; leaves usually deeply lobed . . 12. *O. sinuatum*

Stigmas of long-styled flowers
exserted, of short-styled flowers
not reaching to the bases of the
anthers :
Bracts without a fringe of setae ;
leaves up to 9 × 4 cm., deeply
lobed, ± broadly elliptic in
outline ; large lush herb   .   13. *O. lobatum*
Bracts fringed with setae :
Fruiting pedicels exceeding the
bracts by 4–5 mm., filiform ;
leaves ovate to deltoid, the
base often hastate-cuneate,
usually unlobed ; weak
straggling herb   .   14. *O. atriplicifolium*
Fruiting pedicels scarcely exceed-
ing the bracts ; leaves
lanceolate, elliptic-lanceolate, or ovate-lanceolate :
Leaves pilose, with a central
reddish blotch, shallowly
lobed   .   .   .   15. *O. maculatum*
Leaves ± glabrous, unmarked,
usually shallowly lobed . 16. *O. stuhlmannii*

1. **O. buchananii** (*Dammer*) *Gillett* in K.B. 1953 : 83 (1953). Type :
Nyasaland, without locality, *Buchanan* 550 (K, lecto. !)

A stout perennial (? a bush). Stems glabrous or with scattered papillae.
Ocreae 0·5–2 cm. long, brown, ± glabrous, truncate with a terminal fringe
of 7–9 setae 1–1·3 mm. long. Leaves petiolate, glabrous, large for the genus,
7·5–10·5 × 4·5–5·5 cm., entire or with a slightly wavy edge, broadly obovate
to broadly ovate, apically obtuse (or shortly acuminate) to rounded, basally
long-narrowed and decurrent to the petiole. Petioles 1–2 cm. Inflorescence
elongated, up to 36 cm. long. Bracts 4–5 mm. long, up to 4 cm. apart,
truncate but dorsally produced, with or without a terminal fringe of short
setae up to 1 mm. long. Pedicels glabrous, red, exceeding the bracts by
2·5–3 mm. Flowers 2 (?–4) together. Tepals 3–4 mm. long, oblong. Stamens
almost equalling the tips of the tepals ; anthers orange. Styles not seen.
Very immature fruit with three protuberances on the angles, these probably
being rudimentary prickles. Fig. 5/1 and 2.

TANGANYIKA. Rungwe District : Kasambala village, Jan. 1912, *Stolz* 1090 !
DISTR. **T7** ; Nyasaland
HAB. No information available; about 1200 m.

SYN. *Polygonum buchananii* Dammer in P.O.A. C : 170 (1895)

2. **O. dregeanum** *Meisn.* var. **strictum** (*C. H. Wright*) *R. Grah.* in K.B.
1957 : 167 (1957). Type : Tanganyika, *Scott Elliot* 8365 (K, holo. !, BM, iso. !)

An erect herb with thick, woody, gnarled, reddish-brown, perennial stems
1 cm. or more thick ; and with annual shoots, green becoming red below,
puberulous to thickly pubescent, 20–40 cm. tall, strictly erect or diffuse,
3–4 mm. thick. Ocreae 3–7 mm. long, pubescent, truncate, terminally
membranous and fringed with reddish-brown, bristly setae 6–8 mm. long ;
the leaf inserted in the upper half but often as much as 3 mm. from the apex.
Leaves sessile, erect-patent, linear to narrowly elliptic-lanceolate, 20–30 ×
1–6 (–10) mm., glabrous to thickly pubescent, terminating in a readily

FIG. 5. *OXYGONUM BUCHANANII*—**1**, leafy branch, × 2/3 ; **2**, inflorescence, × 2/3 ; *O. ATRIPLICI-FOLIUM*—**3**, flowering branch, × 2/3 ; **4**, fruit with recurved filiform pedicel, × 2 ; *O. SALICI-FOLIUM*—**5**, flowering branch, × 2/3 ; **6**, fruit, × 2 ; *O. DREGEANUM* var. *STRICTUM*—**7**, flowering branch, × 2/3 ; **8**, fruit, × 2 ; **9**, four diagrams to show relative lengths of styles and filaments in **A** and **B** strongly heterostylous, and **C** and **D** slightly heterostylous flowers, × 2.

deciduous awn 1–2 mm. long. Inflorescence slender, moderately elongated, up to 20 cm. long, the stalk glabrous to thickly pubescent. Bracts 3–4 mm. long, pubescent or glabrous ; terminally membranous or mainly so, truncate but dorsally produced and narrowed into a bristly, pubescent seta 3–4 mm. long with or without 1–2 lateral setae. Pedicels glabrous or pubescent with white hairs, exceeding the bracts by 2–4 mm. ; those of ♂ flowers withering ; those of ⚥ flowers lengthening and thickening. Flowers 1–3 to each bract. Tepals white, 3–4 mm. long, oblong-linear to elliptic-lanceolate. Anthers blue or bluish. Styles of short-styled flowers 1 mm. long ; of long-styled flowers 3–5 mm. long, connate for ⅔ their length. Fruit pubescent, 6 mm. long, bluntly trigonous, lanceolate-ampulliform, broadest below the middle, unarmed, with 2–3 longitudinal ridges down each face and with transverse ridges when ripe. Fig. 5/7–9, p. 29.

TANGANYIKA. "East side [of Lake] Tanganyika," Nov. 1894, *Scott Elliot* 8365 ! ; Kigoma District : Ujiji, Mar. 1939, *Loveridge* 717 ! ; Songea District : about 6 km. W. of Gumbiro, Jan. 1956, *Milne-Redhead & Taylor* 8438 !
DISTR. T4, 8 ; extending southwards and probably found throughout the greater part of south tropical Africa and South Africa
HAB. Sandy ground in *Brachystegia*-woodland, grassland and on disturbed ground ; 600–1500 m.

SYN. *O. delagoense* Kuntze var. *strictum* C. H. Wright in F.T.A. 6 (1) : 100 (1909). Type as *O. dregeanum* var. *strictum*

3. **O. delagoënse** *Kuntze*, Rev. Gen. Pl. 3 : 268 (1898) ; F.T.A. 6 (1) : 100 (1909) ; Fl. Cap. 5 (1) : 461 (1912). Type : Portuguese East Africa, Delagoa Bay, *Kuntze* (NY, holo., K, iso. !)

An erect or ascending herb. Stems up to 1 m. tall, glabrous to pubescent, reddish-brown. Ocreae 4–6 mm. long, pubescent, membranous, truncate, with a terminal fringe of ± light-brown, erect-patent setae 4–5 mm. long. Leaves 2–8 cm. long, often less than 8 mm. broad but up to 1·8 cm., basally decurrent, narrowly elliptic-lanceolate or linear-lanceolate to lanceolate or lanceolate-elliptic, apically acute, mucronate, entire, the margins sinuous, or with 1–3 pairs of forward-directed lobes, sometimes very asymmetrical, puberulent or not on the midrib. Inflorescence elongated, probably up to 30 cm. Bracts 4 mm. long, 1–2·5 (–3) cm. apart, pubescent, with or (commonly) without a fringe of setae. Pedicels 2–4 (–6) together, not exceeding the bracts by more than 3 mm., erect, glabrous or puberulous. Flowers white, 4·75–6 mm. long. Tepals 3–3·75 mm. long. Short styles probably 1·5–2·5 mm. long, free to the base or up to 1 mm. connate. Filaments 1–4·5 mm. long. Fruit 5 (–6) mm. long, pubescent, otherwise smooth, conical, usually armed a little above the base with three small (sometimes scarcely perceptible) ± retrorse teeth or protuberances at the angles.

TANGANYIKA. Probably Dar es Salaam, *Vaughan* 2368 !
DISTR. T?6 ; Portuguese East Africa, Bechuanaland ; also elsewhere in southern and particularly in coastal SE. Africa
HAB. Roadsides, disturbed and weedy ground, and in crops (maize)

NOTE. The above record is based on a specimen in the British Museum herbarium which is probably this species, but only vegetative characters are available for identification owing to the absence of all but very immature fruit.

4. **O. sagittatum** *R. Grah.*, K.B. 1957 : 158 (1957). Type : Kenya, Northern Frontier Province, Dandu, *Gillett* 12694 (K, holo. !)

A rather hoary, decumbent herb, probably perennial. Stems greenish-brown, densely pubescent with white hairs, glabrescent with age. Ocreae often crowded and obscuring the stem, membranous, densely pubescent,

0·7–1·2 cm. long, the leaf inserted a little above the middle, terminally truncate and fringed with pubescent, red setae 7–10 mm. long. Leaves petiolate, narrow-lanceolate, up to 2·7 × 0·6 cm., the margin entire or with 1 small sagittate lobe half way along, densely and closely hairy on both sides, very acute with an extended mucro, basally sagittate with lobes up to 6 mm. long, and below the lobes often shortly cuneate to the petiole, the margins revolute, and the midrib very prominent below. Petiole 0·6–1·2 cm. long. Inflorescence ± stout, elongated, often arcuate or gently so, 6–18 cm. long. Bracts 2–4 mm. long, with a dorsal seta 4 mm. and lateral setae 1–2 mm. long. Pedicels just exceeding the bracts. Flowers 1–3 together, white flushed with pink, ± 5 mm. long ; tepals oblong ; long styles 4 mm. long, connate for 1·5 mm. (short styles not seen) ; stigmas yellow. Fruit (immature) 6 mm. long, fusiform, densely pubescent, with three prickles 1 mm. long, spreading from the centre. Fig. 6/8, p. 35.

KENYA. Northern Frontier Province : Dandu, Apr. 1952, *Gillett* 12694 !
DISTR. **K**1 ; not known beyond this single record ; it should be sought for in the Kenya-Ethiopia boundary area
HAB. In *Acacia-Commiphora* bushland ; 1100 m.

5. **O. salicifolium** *Dammer* in P.O.A. C : 171 (1895) ; F.T.A. 6 (1) : 102 (1909). Type : Tanganyika, Tanga District, Duga, *Holst* 3186 (B, lecto. !, K, iso.-lecto. !)

A creeping herb, with ascending or prostrate branches. Creeping stems woody, reddish-brown, subglabrous to pilose ; shoots up to 35 cm. long, pubescent. Ocreae 0·4–1 cm. long, membranous, thinly or thickly covered with whitish hairs, terminally truncate and fringed with red setae 6–9 mm. long, the leaf inserted usually in the centre or slightly below. Leaves sessile, narrowly linear-lanceolate, broadest towards the base, (30–) 50–60 (–70) × (2–) 3–8 (–10) mm., not or only slightly sinuous on the margin, apically acute, often mucronate, basally narrowed, the midrib prominent below, whitish, almost glabrous on both faces, or pilose mainly on the margins and midrib. Inflorescence slender, 12–20 cm. long. Bracts 3–3·5 mm. long, glandular, usually glabrous, truncate but produced dorsally to an acute setose apex, with a terminal fringe of 3–5 setae which may exceed the length of the bract. Tepals 4 mm. long (6 mm. *fide* Dammer) ; linear-lanceolate, the tube very short ; filaments 3·5–4 mm. ; short-styled flowers with styles 1 mm. long, free almost to the base ; long-styled flowers with styles ± 4 mm. long, exserted. Fruit pubescent, 10 mm. long, with three prickles 3–3·5 mm. long, spreading horizontally (or slightly ascending) from the centre or from up to 1 mm. above it. Fig. 5/5 and 6, p. 29.

KENYA. Machakos District : Lukenya, Sept. 1954, *Bally* 9851 ! ; Mombasa District : Changamwe, Mar. 1902, *Kassner* 262 ! ; Kilifi District : Arabuko, May 1929, *R. M. Graham* 2138 !
TANGANYIKA. Tanga District : Moa, Aug. 1953, *Drummond & Hemsley* 3634 ! ; Mpwapwa, Feb. 1919, *Hornby* 482 !
DISTR. **K**4, 7 ; **T**1, 3, 5, 7 ; not known elsewhere
HAB. Grasslands and disturbed ground ; 0–1560 m.

NOTE. *Bally* 9851 is apparently a form with more thickly pilose leaves, both this and its somewhat starved state being perhaps due to inclement conditions during the growth of the shoots.

6. **Oxygonum subfastigiatum** *R. Grah.* in K.B. 1957 : 156 (1957). Type : Tanganyika, between Orero & Kilwa Kivinje, *Braun* 1305 (K, holo. !, EA, iso. !)

A perennial much-branched herb. Stems woody, reddish-brown, bearing many congested, persistent ocreae from previous seasons' growth, pubescent or becoming glabrous. Ocreae 7 mm. long on the old stems, 2·5–7 mm. long on the new growth, membranous, glabrous or pubescent, terminally truncate and fringed with setae 5–9 mm. long. Leaves sessile, glabrous, congested towards the base of the current season's growth, 25–35 × 2–2·5 mm., ± erect, subfastigiate, linear, narrowed to each end, apically mucronate, the midrib prominent below, rather revolute at the margins. Inflorescence slender, up to ± 15 cm. long. Bracts glabrous, truncate but produced dorsally into a long point, terminally fringed with setae up to 3 mm. long. Flowers 2–3 from each bract. Pedicels long, 7–13 mm., glabrous or puberulent, spreading becoming deflexed. Flowers not seen. Very immature fruit apparently prickly. Fig. 6/5, p. 35.

TANGANYIKA. Kilwa District : between Orero and Kilwa Kivinje, Feb. 1906, *Braun* 1305 !
DISTR. T8 ; unknown beyond this single record
HAB. Sandy soil

NOTE. In the woody, perennial stems with persistent, congested ocreae, this species suggests the conditions of its habitat in that it is perhaps covered by successive layers of sand as the seasons go by.

7. O. schliebenii *Mildbr.* in N.B.G.B. 14 : 103 (1938). Type : Tanganyika, Lindi District, Lukuledi Valley, *Schlieben* 6475 (B, holo. !, BM, iso. !)

A straggling, perhaps climbing, perennial. Stems decumbent, rather woody towards the base, varying from densely pubescent with brown hairs to glabrous. Ocreae 0·5–0·7 mm. long, ± densely pubescent, membranous, terminally truncate with a fringe of rather wavy setae of similar length. Leaves sessile, rather like those of *O. salicifolium*, but broader, elliptic, lanceolate, or ovate-elliptic, 5–6 × 1·1–1·8 cm., very acute, terminally mucronate, marginally entire or slightly sinuous, the midrib of lower face often pale brown, glabrous or thinly pubescent on the midrib below. Inflorescence slender, elongated, somewhat arcuate, 50 or more cm. long. Bracts glabrous, fringed with setae 1–1·25 mm. long. Pedicels pubescent ; those of ♂ flowers 3–6 mm. long ; those of ♀ flowers noticeably very long, 17–24 mm. Tepals of long-styled flowers 2 mm. long, linear-lanceolate ; styles (only long styles seen) 3·75 mm. long, connate for 1·5 mm. Fruit fusiform, densely pubescent when young, later glabrescent, up to 6 mm. long, armed at the centre with 3 spreading prickles, 3 mm. long, on the angles and with 1–3 intermediate subsidiary prickles up to 2 mm. long on each intervening face.

TANGANYIKA. Iringa District : Usagara, Mazombe, July 1936, *Ward* U7 !; Makunga, Feb. 1909, *Zimmermann* 55 ! ; Kilwa, *Kirk* !
DISTR. T7, 8 ; no certain records apart from the above and the type, but likely to be more widespread
HAB. " Roadsides " ; perhaps a plant of disturbed ground ; 0–1400 m.

NOTE. The *Kirk* specimen has lanceolate, more sinuous leaves, while those of *Ward* U7 are broader, ± ovate-elliptic and the veins of the undersurface are more pronounced. It is possible that another, unrecognized species is involved, but the evidence for this is not adequate.

8. O. leptopus *Mildbr.* in N.B.G.B. 11 : 810 (1933). Type : Tanganyika, Ulanga District, *Schlieben* 2324 (B, holo. !, BM, iso. !, HBG, iso. !)

A straggling perennial with a thick, perennial, gnarled, woody rootstock. Annual stems reddish, densely pubescent, becoming ± glabrous above. Ocreae ± 5 mm. long, pubescent, membranous, truncate with a terminal fringe of setae 5–7 mm. long. Leaves shortly but clearly petiolate, ovate to

ovate-deltoid, marginally entire, 3·2–3·5 × 1·5–2·0 cm., apically subacute
with the midrib excurrent as a mucro, basally rounded or ± so to the petiole,
densely pustular, pubescent only on the midribs of both faces.   Petioles
densely pubescent, 2 mm. long. Inflorescence very long, slender (50–) 70–
125 cm. long, arcuate. Bracts 3–4 mm. long, ± glabrous, up to 2·8 (–3) cm.
apart, sometimes in semi-adjacent pairs, fringed with setae up to 2 mm.
long.   Pedicels 5–9 to each bract ; those of ☿ flowers very long, 1·5–2·5
(–3) cm., and usually only 1–2 per bract ; of ♂ flowers 3–5 mm. long, filiform.
Tepals ovate-oblong or oblong, 2·75 mm. long (2·5–3 mm. ; styles clearly
exceeding the stamens, *fide* Mildbr.). Short styles 1 mm. long, connate
0·4 mm. (Immature fruit fusiform-trigonous, villous, with three small erect
prickles, *fide* Mildbr.)

TANGANYIKA. Ulanga District : between the Mahenge plateau and the confluence of
   the Kilombero and Luwegu Rivers with the R. Rufiji, June 1932, *Schlieben* 2324 !
DISTR. T6 ; unknown apart from the type locality
HAB. Dry, sandy woodland, frequent in low grass ; about 400 m.

NOTE. Despite the absence of knowledge concerning fruit, this is immediately told from
   all other African spp. by the entire, ± ovate leaves with their rounded or near-
   rounded bases, and the very short but distinct petioles.

9. **O. hirtum** *Peter* in F.D.O.-A. 2 : 188 (1932) & anh. : 18, t.22/1 (1938).
Type : Tanganyika, Pare District, near Same, *Peter* 11883 (B, holo.†)

A subglaucous, little-branched herb, minutely pubescent throughout,
± 60 cm. tall. Stems slender ; internodes 3·5–4 cm. Ocreae 7 mm. long, not
tightly appressed, terminally fringed with setae 7–9 mm. long. Leaves
3–4 × 1·5–2 cm., elliptic-ovate but deeply incised and irregularly dentate,
attenuate at each end, shortly puberulous. Petioles up to 15 mm. long.
Inflorescence elongate. Bracts 3·5–4 cm. apart. Pedicels 5–8 mm. long.
Flowers 4–5 mm. long ; tepals oblong, pubescent, narrowly winged at the
back. Fruit 8 mm. long, with 3 horizontal prickles 3 mm. long.

TANGANYIKA. Pare District : near Same, *Peter* 11883
DISTR. T3 ; not known elsewhere
HAB. Unknown; about 900 m.

NOTE. No specimens seen. From the figure, this species would suggest a hirtellous
   form of *O. sinuatum*, but with unusually long fruiting pedicels. Further collecting is
   very desirable in the type neighbourhood.

10. **O. ellipticum** *R. Grah.* in K.B. 1957 : 156 (1957). Type : Tanganyika,
Mpanda District, Lake Katavi, *Bullock* 2333 (K, holo. !)

A slender, semi-prostrate, spreading herb. Stems green becoming brown,
puberulous, with long internodes 6–9 cm. apart. Ocreae 0·8–1·1 cm. long,
appressed when young, becoming more open with age, puberulous, truncate,
with a terminal fringe of setae 4–4·25 mm. long, the leaf inserted in the upper
half but often near the middle. Leaves petiolate, elliptic-lanceolate to ovate-
elliptic, 5–5·5 × 1·5–2 cm., broadest at or a little below the middle, apically
acuminate with a short terminal mucro, gradually narrowed and decurrent
to the petiole, puberulous on the veins of both sides. Petioles 1–1·5 cm.
long. Inflorescence slender, up to 30 cm. long. Bracts 4–5 mm. long,
1–5 cm. apart, truncate but produced dorsally into a seta 3–4 mm. long, with
1–2 pairs of shorter, lateral setae. Pedicels glabrous or minutely puberulous,
exceeding the bracts by 3–4 mm. Flowers 1–4 to each bract. Tepals 4–5 mm.
long, linear-oblong to obovate. Styles yellow, 5 mm. long, connate for 2 mm.
(only long styles seen). Fruit fusiform, glabrous or with scattered papillae,
7–8 mm. long and almost as wide, with three central spreading prickles 2 mm.

long on the angles, with or without intermediate, smaller prickles on the faces, the angles also sometimes winged and in the lower half sometimes with 1–2 small teeth on the angles which may not project beyond the wing. Fig. 6/6 and 7.

TANGANYIKA. Mpanda District : Lake Katavi, Jan. 1950, *Bullock* 2333 !
DISTR. **T4** ; not certainly recorded elsewhere
HAB. Sandy littoral of the lake ; 1050 m.

11. **O. magdalenae** *Peter* in F.D.O.-A. 2 : 186 (1932). Type : Tanganyika, near Iringa, *Magdalene Prince* (B, holo. !)

A stout, erect herb up to 50 cm. tall. Stems brownish red, thickly hairy below with short brown hairs, becoming less so above, or glabrous throughout. Ocreae 7–8 mm. long, hairy as the stem, truncate, terminally fringed with setae 5–6 mm. long, the leaf inserted a little above the middle. Leaves sessile to shortly petiolate, elliptic, 35–45 × 9–13 mm., acute with a terminal mucro 0·5 mm. long, ± evenly narrowed to each end or more shortly so to the base, shortly and varyingly hairy on both surfaces but especially thickly on the midrib and veins below and on the margin, sometimes glabrous throughout. Petioles 0–6 mm. long. Inflorescence stout, up to and mainly 2 mm. thick (sicc.). Bracts 4–6 mm. long, up to 7 cm. apart, truncate but dorsally produced into a narrow-lanceolate point 4 mm. long with 2 lateral setae on each side. Pedicels 2–4 together, those of the fruit exceeding the bracts by 5 mm. Flowers showy, 7–9 (? more) mm. across. Short-styled flowers : tepals 3 mm. long, ovate to oblong ; styles 2 (–1·5) mm. long, connate for just under half their length (or free to the base), reddish. Long-styled flowers : tepals 4 mm., styles 4 mm. long, connate for 1 mm. Fruit pubescent (glabrous *fide* Peter), fusiform, 9–10 m. long, with three prickles 3–4 mm. long spreading horizontally from the centre.

TANGANYIKA. Iringa, Feb. 1932, *Lynes* I.g. 130 ! ; Iringa District, near Kalinga, *Ward* K1 ! ; Songea, shortly W. of the town, Feb. 1956, *Milne-Redhead & Taylor* 8497 !
DISTR. **T7, 8** ; not known elsewhere
HAB. *Brachystegia*-woodland and probably upland grassland ; 1000–1600 m.

12. **O. sinuatum** (*Meisn.*) *Dammer* in P.O.A. C : 170 (1895) ; F.C.B. 1 : 406 (1948). Type : Ethiopia, Chachito, *Schimper* 264 (K, lecto. !)

A diffuse, decumbent or erect, weedy annual. Stems glabrous to pubescent, green to reddish brown. Ocreae reddish, up to 5·5 mm. long, truncate, usually fringed with setae, the leaf inserted in the upper half and often near the apex. Leaves petiolate, in outline ovate, ovate-elliptic, ovate-lanceolate or elliptic-lanceolate, but usually deeply incised with rounded or acute lobes, often panduriform or lyrate, narrowed to each end, commonly 4 × 1·5 cm., but up to 6 × 2·7 cm., usually pustular on the undersurface, otherwise glabrous or shortly pilose on the veins. Petioles 1–2 cm. long. Inflorescence variably elongated, up to 28 cm. long (or more ?), the stalk up to 2 mm. thick. Bracts up to 6 cm. apart, fringed with setae. Pedicels of ♀ flowers stumpy, scarcely filiform, exceeding the bracts by 0–1 (–3) mm. ; those of ♂ flowers ± filiform, withering. Flowers white or pink, slightly heterostylous, i.e. long-styled stigmas reaching only to the tips of the anthers or to slightly beyond, and short-styled stigmas reaching to or nearly to the anther bases (see Fig. 5/9). Tepals ovate-elliptic, 2·5–3 mm. long ; tube 1–1·5 mm. long. Fruit fusiform, (1–) 2–4 to each bract, erect or spreading but scarcely pendulous, 5·0–6·5 mm. long when mature, pubescent or papillose, more rarely glabrous, with three spreading prickles 1·5–1·75 (–2) mm. long arising

FIG. 6. *OXYGONUM MACULATUM*—**1,** flowering branch showing spotted leaves, × 2/3 ; *O. LOBATUM* —**2, 3,** flowering branch, × 2/3 ; **4,** fruit, × 2 ; *O. SUBFASTIGIATUM*—**5,** flowering branch, × 2/3 ; *O. ELLIPTICUM*—**6,** flowering branch with inflorescence diagrammatically deflexed, × 2/3 ; **7,** fruit × 2 ; *O. SAGITTATUM*—**8,** flowering branch, × 2/3.

on the angles centrally or more often from just below the middle (rarely with 3 smaller, intervening prickles on the faces).

UGANDA. W. Nile District : Warr, Attiak, Apr. 1940, *Eggeling* 3912 ! ; Ankole District : Mbarara, Mar. 1939, *Purseglove* 601 ! ; Busoga District : Jinja, Ripon Falls, July 1914, *Dummer* 23 !
KENYA. Naivasha District : Kedong, Mt. Margaret Estate, June 1940, *Bally* 907 ! ; Kiambu, Aug. 1938, *Leakey* 67 !; Masai District : Ngong Hills, Aug. 1951, *Bally* 8025 !
TANGANYIKA. Old Shinyanga, Jan. 1940, *Welch* 9 ! ; Masai District : Ngorongoro Crater, Apr. 1941, *Bally* 2262 ! ; Mbeya District : Poroto Mts., Mar. 1932, *St. Clair-Thompson* 776 !
DISTR. U1–4 ; K2–4, 6, 7 ; T1–8 ; eastern Africa from Sudan southwards, Belgian Congo
HAB. Cultivated and other ground suited to weed growth ; 0–2100 m.

SYN. [*Ceratogonum atriplicifolium* sensu A. Rich., Tent. Fl. Abyss. 2 : 231 (1851) pro parte, ex descr., *non* Meisn.]
  Ceratogonum sinuatum Meisn. in DC., Prodr. 14 : 40 (1856)
  Ceratogonum cordofanum Meisn. in DC., Prodr. 14 : 39 (1856). Type : Nubia, *Kotschy* 117 (K, lecto. !)
  Oxygonum elongatum Dammer in P.O.A. C : 170 (1895), quoad spec. cit. *Stuhlmann* 288 !
  O. atriplicifolium var. *sinuatum* (Meisn.) Baker in F.T.A. 6 (1) : 101 (1909)

NOTE. This species is distinguished from *O. atriplicifolium*, with which it is sometimes confounded, by its habit ; stouter stems and racemes ; usually deeply lobed leaves ; short, ± erect fruiting pedicels ; and by the less pronounced degree of heterostyly ; and from *O. stuhlmannii*, which it often greatly resembles particularly when it has markedly elongated racemes, by the far stronger heterostyly of the latter species.

13. **O. lobatum** *R. Grah.* in K.B. 1957 : 159 (1957). Type : Tanganyika, about 14·5 km. from Morogoro on Dakawa road, *Drummond & Hemsley* 1775 (K, holo. !)

A stout, ± glabrous, lush, perennial herb 50 cm. or more tall. Stems 3–5 mm. thick, reddish-brown. Ocreae pustular, brown, papery, glabrous, readily tearing, truncate, fringed with pubescent setae 4–5 mm. long, the leaf inserted in the upper part. Internodes 4–8 cm. Leaves petioled, minutely pustular, 6·5–9 × 2·5–4 cm., ± ovate or ovate-elliptic in outline but strongly and irregularly lobed, apically subacute with a membranous mucro, basally narrowed and decurrent (often unequally) to the petiole, glabrous above and below. Petioles brown, 1–2 cm. long. Inflorescence stout, elongated, 35–40 cm. long, the racemes single or in pairs from the lowest bract. Bracts 4–6 mm. long, up to 4 cm. apart, foliaceous and terminally membranous, cup-shaped and rather abruptly widening at the base, truncate but produced dorsally into a lanceolate point ± 2 mm. long, without a terminal fringe of setae. Pedicels exceeding the bracts by 1–2 mm. Tepals green outside, cream inside, 4 mm. long, oblong with a short, ± hooded, acute apex ; filaments 2 mm. long in long-styled flowers, 4 mm. in short-styled flowers, white, filiform ; anthers 0·75–1 mm. long, linear ; long styles 4 mm. long, connate for 2·0 mm. ; short styles 1·5 (–2) mm. long. Fruit 10 × 6 mm., trigonous, fusiform-conical, broadest towards the base, glabrous, with three short, slightly reflexed prickles 1–1·5 mm. long, spreading from below the middle when immature but distinctly from nearer the base when ripe. Fig. 6/2–4, p. 35.

TANGANYIKA. Morogoro District : about 14·5 km. from Morogoro on Dakawa road, Mar. 1953, *Drummond & Hemsley* 1775 ! and Uluguru Mts., Dec. 1935, *E. M. Bruce* 375 !
DISTR. T6 ; unknown apart from these two records
HAB. " Scattered thornbush with poorly developed ground cover " ; also as a weed of native cultivation; 600 m.

14. **O. atriplicifolium** (*Meisn.*) *Martel.*, Fl. Bogos. : 69 (1886) ; F.T.A. 6 (1) : 101 (1909). Type : Calcutta Botanic Garden, origin unknown (presumably Africa) *Wallich* 1719 (K–W, lecto. !)

A slender, much branched, straggling and trailing herb. Stems weak, pubescent mainly down one side, up to 1 m. long or more. Ocreae up to 1 cm. long, light brown, membranous, pubescent, not closely appressed, truncate with a terminal fringe of setae which may exceed the length of the tube ; the leaf inserted in the upper half. Leaves small, petiolate, commonly 2–3 × 1·5–1·8 cm., deltoid, deltoid-ovate, or deltoid-lanceolate, marginally entire or slightly uneven, but not deeply lobed, apically very acute, the apex sometimes aristate, basally truncate to cuneate (rarely sagittate), sometimes ± hastate, glabrous above, the margin and veins of the under-surface pubescent, the lamina often pustular below. Petioles pubescent, 1–1·5 cm. long. Inflorescence slender, leafless, up to 30 cm. or more long, the stalk less than 1 mm. thick. Bracts up to 25 mm. apart below, 3–4 mm. long, pubescent and fringed as the ocreae. Pedicels filiform, pubescent ; those of ♀ flowers lengthening, exceeding the bracts by 4–6 mm., becoming patent and later reflexing ; those of ♂ flowers neither lengthening nor reflexing but withering, not exceeding the bracts by more than ± 3 mm. Flowers white or greenish outside, strongly heterostylous ; long-styled flowers with broadly linear or ovate tepals, 1·5 mm. long ; styles 2 mm., connate 0·75 mm. (short-styled flowers not seen) (♂ flowers with 5 narrow, oblong, obtuse, flat tepals ; ♀ flowers with 6 tepals, the inner 3 petaloid, the outer 3 keeled, *fide* Meisner). Fruit fusiform, 5–6·5 mm. long, 1 (–2) to each bract, pendulous, glabrous or pubescent (sometimes on the same plant), bearing 3 spreading prickles ± 1 mm. long at the centre or slightly below. Fig. 5/3 and 4, p. 29.

KENYA. Mombasa District : Mombasa, Nov. 1951, *Bogdan* 3298 ! ; Mombasa to Malindi road at 35 km., Dec. 1953, *Verdcourt* 1087 ! ; Malindi, Oct. 1951, *Tweedie* 942 !
DISTR. **K7** ; Somalia and Portuguese East Africa ; also recorded from Madagascar
HAB. Hedgerows, cultivated and waste ground ; altitude range uncertain, but descending to sea level

SYN. *Ceratogonum atriplicifolium* Meisn. in Wall., Pl. As. Rar. 3 : 63 (1832) and in DC., Prodr. 14 : 39 (1856).
Polygonum owenii Bojer, Ann. Sc. Nat., sér 2, 4 : 267 (1835). Type : Kenya, Mombasa, *Owen* (location unknown)
O. somalense Chiov., Miss. Stef.-Paoli Somal. Ital. 1 : 152 (1916). Type : Somalia, by R. Uebi Scebeli N. of Mogadishu, *Paoli* 1323 (FI, holo. !)
Oxygonum denhardtii De Wild., Pl. Bequaert. 4 : 313 (1928). Type : Kenya, Kilifi District, Takaungu, *F. Thomas* II, 64 (BR, holo. !)
Oxygonum fagopyroïdes Peter in F.D.O.-A. 2 : 186 (1932). Type : Kenya, Kilifii District, Takaungu, *F. Thomas* II, 64 (B, lecto. !)

15. **O. maculatum** *R. Grah.* in K.B. 1957: 163 (1957). Type : Kenya, Machakos District, *V. G. L. van Someren* 1599 (K, holo. !)

Probably perennial herb, much branched, generally whitish-green due to an abundance of white hairs. Stem ? decumbent, or creeping with erect annual branches, 30–45 cm. tall or more, woody and glabrous below, covered with spreading or reflexed white hairs above, apparently sometimes glabrous down one side. Ocreae 6 mm. long, pilose, the leaf inserted at the middle or rather below, fringed with brown setae of similar length. Leaves shortly petiolate, lanceolate, 2–4 × 0·2–1 cm., the margin crisped and uneven, usually red, apically acute, long decurrent at the base, thickly pilose on both faces or glabrescent above, centrally marked on the upper surface with a lanceolate, purple or red blotch. Inflorescence elongated, erect or slightly arcuate, up to 20 cm. long (or more ?). Bracts 4–5 mm. long, fringed with short setae. Pedicels erect, 3–3·5 mm. long, filiform, glabrous or thinly pilose. Flowers 2–3 together, pale pink, clearly heterostylous. Tepals punctate, 4 mm. long, linear-lanceolate ; styles 3 mm. long, connate for 1 mm. (long styles only seen). Fruit fusiform, 6–6·25 mm. long, green, pilose, with three slightly reflexing or spreading prickles 1 mm. long, rising from the centre or a little below. Fig. 6/1, p. 35.

KENYA. Machakos District : unlocalized, Dec. 1931, *V.G.L. van Someren* 1599 ! and
    Kapiti Plains, at foot of Mwami Hill, June 1957, *Bally* 11525 !
DISTR. **K4** ; not elsewhere recorded
HAB. Upland grassland ; the plant is said to grow in masses and to flower profusely ;
    1710 m.

16. **O. stuhlmannii** *Dammer* in P.O.A. C : 171 (1895) ; F.T.A, 6 (1) : 102 (1909). Type : Tanganyika, Mwanza, *Stuhlmann* 4628 (B, holo. !, BM, iso. !)

? Perennial herb, probably with a trailing woody base, and with ascending, reddish, puberulous stems. Ocreae 5–8 mm. long, membranous, puberulous, truncate with a terminal fringe of rather wavy setae ± 5 mm. long. Leaves petiolate, up to 5 × 1·2 cm., lanceolate, elliptic-lanceolate, or subdeltoid, the margins ± entire or more usually shallowly sinuous, very acute long narrowed to the base, glabrous, pustular on the undersurface. Petioles rather long, up to 1 cm. in length. Inflorescence elongated, rather stout, up to ± 35 cm. long, straight or slightly arcuate. Bracts 2·5–4 mm. long, puberulous to glabrous, fringed with 3–5 setae ± 2 mm. long. Pedicels some-what exceeding the bracts. Flowers strongly heterostylous. Long styles 3–3·5 mm. long, free almost to the base, strongly exserted. Short styles 1 mm. long or slightly less. Tepals ovate-elliptic, 2 mm. long (5 mm. in long-styled flowers, *fide* Dammer). Fruit 5 mm. long, fusiform, thinly papillose, with 3 spreading prickles a little below the middle.

KENYA. Elgeyo District : Elgeyo Escarpment, *Harger* ! ; Naivasha District : Gilgil
    River, *Scott Elliot* 6651 ! ; Teita District : Voi Station, Dec. 1949, *Jeffery* 714 !
TANGANYIKA. Mwanza District : Ouma, Oct. 1951, *Tanner* 334 ! ; Handeni District :
    Kideliko, Apr. 1954, *Faulkner* 1431 ! ; Tabora, Mar. 1937, *Lindeman* 312 !
DISTR. **K3** ; **T1–4** ; Belgian Congo, also probably in Nyasaland and Rhodesia
HAB. Cultivated or disturbed ground ; from 2100 m. probably nearly to sea level

SYN. *O. fasciculatum* C. H. Wright in F.T.A. 6 (1) : 102 (Mar. 1909) ; K.B. 1909 :
        186 (May 1909). Type : Kenya, Naivasha District, Gilgil River, *Scott Elliot*
        6651 (K, lecto. !, BM, iso-lecto. !)
    ?*O. baumii* C. H. Wright in F.T.A. 6 (1) : 103 (1909). Type : Angola, Kuito,
        Baum 532 (K, holo. !—spec. immature)
    O. humbertii Robyns in B.S.B.B. 17 : 160 (1944). Type : Belgian Congo, Kivu,
        Humbert 7284 (BR, holo. !)

NOTE. *O. stuhlmannii* appears to be a species of considerable variability in leaf characters
    both as regards shape and degree of hairiness.

7. **ANTIGONON**
Endl., Gen. Pl. : 310 (1837)

A. leptopus *Hook. & Arn.*, Bot. Beech. Voy. : 308, t. 69 (1839)

A shrubby climber with tendrilous branches, cordate leaves with petioles basally ± amplexicaul but without ocreae, with 6 (3 + 3) pink or white, accrescent, papery, reticulately veined tepals up to about 2 cm. long, and with nuts 8–9 mm. long, very acute and sharply trigonous in the upper half.

KENYA. Mombasa District : Nyali Beach, Sept. 1952, *Starzenska* 6 !
TANGANYIKA. Morogoro District : Wami Plains, Jan. 1932, *Wallace* 258 ! (? planted)
    & near Morogoro, *Peter* 32013 ; Dar es Salaam, *Peter* 39414
DISTR. **K7**, **T6** ; a native of Mexico, introduced into our area.
HAB. Occurs locally as an escape from cultivation.

INDEX TO POLYGONACEAE